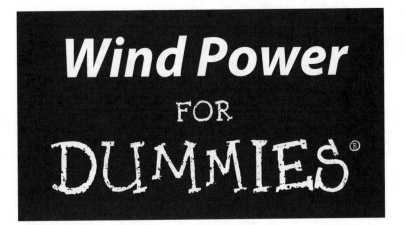

Wind Power
FOR
DUMMIES®

by Ian Woofenden

WILEY

Wiley Publishing, Inc.

Wind Power For Dummies®

Published by
Wiley Publishing, Inc.
111 River St.
Hoboken, NJ 07030-5774
www.wiley.com

About the Author

Ian Woofenden began exploring renewable energy as a preteen growing up in the Midwest, where he built a solar greenhouse, used a clothesline, and walked, ran, and bicycled extensively. Later, when he was a teenager in New England, his parents had a solar hot water system, and he read widely on sustainability, renewable energy, and country living.

After marrying his college sweetheart and going on a 1,200-mile honeymoon on a tandem bicycle, Ian lived in a tipi in Maine and on an island in Lake Michigan before settling on an island in the inland waters between mainland Washington state and Vancouver Island, Canada. There, he and his wife bought 10 acres of off-grid property and began an experimental life with country living and renewable energy that has now spanned almost three decades.

Raising a large family off-grid had many lessons to teach, and running a production woodcraft business for 14 years was challenging and instructive, too. Starting with an automobile battery and a few taillight bulbs, Ian's renewable energy systems have grown to include three wind generators, multiple solar-electric systems, two solar hot water systems, wood heating, extensive gardens and orchards, and again, clotheslines. Ian's favorite mode of transportation is a bicycle, and a solar recumbent tandem trike may be in his future.

Ian comes to renewable energy first and foremost as a user and abuser, an experimenter and active learner. In the early 1990s, Ian decided to pursue this passion as a career and began to take workshops offered in Colorado by Solar Energy International (SEI). A year later, he was coordinating workshops in the Northwest for SEI. Three years later, he landed a job as an editor with *Home Power* magazine, his all-time favorite publication. Today, Ian is one of the senior editors at *Home Power* and is Northwest and Costa Rica Coordinator for SEI, organizing and co-teaching 8 to 10 weeks of workshops per year.

Ian is author of numerous articles on wind energy and other renewable energy topics for *Home Power* and other publications, and he is one of the supporting coauthors of *Power from the Wind* by Dan Chiras. He also teaches wind-energy workshops for other organizations and does private consulting for individuals, businesses, and organizations. He particularly enjoys teaching and consulting in Central America, where he spends several weeks each winter.

With his family mostly grown, Ian is excited to see some of his kids involved in renewable energy and environmental education. His family homestead is still an experimental lab where new products are tested and new lessons are learned. If he hasn't already overcommitted himself, Ian likes to correspond with readers at ian.woofenden@mindspring.com.

Dedication

This book is dedicated to my seven children, who have the capacity to change the world they live in by changing themselves and the way they live; my wife, who has lived with my wind-energy addiction for lo these 31 years; my parents, who taught me by example to think for myself, read, write, and care about the important things in life; and my many friends, supporters, readers, and students, who have discussed, laughed, cried, critiqued, and ranted with me over the years.

Author's Acknowledgments

As an editor, I have been intimately involved in shaping others' writing, and I know from experience that no one is an island. Anything produced is the result of collaboration on many levels. I stand on the shoulders of others who have been working in the field much longer than I and on those I have shared space with in foundation holes and on top of towers in 25 mph winds.

In particular, I'd like to recognize the following:

Hugh Piggott has been a source of much information, experience, and perspective for many years. I appreciate not only his technical savvy but also his humility and generosity. Having him as technical reviewer on this book is another high point in a long trail as friends and colleagues.

Mick Sagrillo is gradually leaving huge shoes to fill, and I appreciate his willingness to share his knowledge, his recommendations, and his time. As mutual thorns in each other's sides, being around the two of us has been said to be "worth the price of admission."

Paul Gipe is a model of straight-up journalism, and he knows more about more wind generators — large and small — than anyone else I know. He calls a spade a spade, and for that I have great respect.

Other renewable energy colleagues who have helped in my education are my earliest renewable energy guru Windy Dankoff, Christopher Freitas, Richard Perez, Joe Schwartz, Michael Welch, Johnny Weiss, Robert Preus, Dan New, Ed Kennell, Mike Klemen, Dan Fink, Dan Bartmann, Randy Brooks, Kelly Keilwitz, Rose Woofenden, Bill Hoffer, Brent Summerville, E. H. Roy, Roy Butler, Steve Wilke, Tod Hanley, Darren Emmons, Chuck Marken, Megan Amsler, Conrad Geyser, Tom Wineman, Randy Richmond, Victor Creazzi, Eric Eggleston, Jason Lerner, Dana Brandt, the crews at *Home Power* magazine and Solar Energy International, and many others.

My island neighbors and renewable energy users have played a strong role in my renewable energy education. I especially appreciate wind-energy system owners Holly and Kevin Green, John Meyer and Lisa Kennan-Meyer, Frank and Deb Dehn, Blake and Nancy Rankin, and the Anderson family.

My editors at Wiley, Mike Baker, Danielle Voirol, Megan Knoll, and especially Georgette Beatty, have been professional, patient, and insightful. They have formed my rough book into a saleable creature. The many people behind the scenes at Wiley also have my appreciation.

I am blessed with many friends inside and outside of the renewable energy world who give me support, feedback, and encouragement in following my chosen paths in life, with its struggles and triumphs. In particular, I'd like to mention Clay Eals, Andy Gladish, Juby Fouts and clan, Heather Isles, my sister Laura, Susan Miller, Doug Moser, and Steve Dyck, among many, many others.

My immediate and extended family has been an inspiration and a blessing. Several writers are among them, including my father and mother; my favorite and only surviving uncle, George; my brother Lee; and my terribly missed father-in-law, Dave Gladish. My children and future grandchildren are a big part of my inspiration. My hope is that this book reduces wasted resources, time, and money and helps people use one of our abundant natural resources more wisely.

While I'm appreciative of all that these many people have contributed to my life and ultimately this book, responsibility for errors, missing info, and my ever-present personal biases is mine, all mine. Life is imperfect and short; take what you like and leave the rest.

Publisher's Acknowledgments

We're proud of this book; please send us your comments through our Dummies online registration form located at http://dummies.custhelp.com. For other comments, please contact our Customer Care Department within the U.S. at 877-762-2974, outside the U.S. at 317-572-3993, or fax 317-572-4002.

Some of the people who helped bring this book to market include the following:

Acquisitions, Editorial, and Media Development

Senior Project Editor: Georgette Beatty

Acquisitions Editor: Mike Baker

Senior Copy Editor: Danielle Voirol

Copy Editor: Megan Knoll

Assistant Editor: Erin Calligan Mooney

Editorial Program Coordinator: Joe Niesen

Technical Editor: Hugh Piggott

Editorial Manager: Michelle Hacker

Editorial Assistant: Jennette ElNaggar

Art Coordinator: Alicia B. South

Cover Photo: Ian Woofenden

Cartoons: Rich Tennant
(www.the5thwave.com)

Composition Services

Project Coordinator: Sheree Montgomery

Layout and Graphics: Joyce Haughey, Melissa K. Jester, Mark Pinto, Melissa K. Smith

Special Art: Thomas Brucker, Precision Graphics (precisiongraphics.com)

Proofreaders: Caitie Copple, Evelyn C. Gibson

Indexer: Estalita Slivoskey

Special Help

Elizabeth Rea

Publishing and Editorial for Consumer Dummies

Diane Graves Steele, Vice President and Publisher, Consumer Dummies

Kristin Ferguson-Wagstaffe, Product Development Director, Consumer Dummies

Ensley Eikenburg, Associate Publisher, Travel

Kelly Regan, Editorial Director, Travel

Publishing for Technology Dummies

Andy Cummings, Vice President and Publisher, Dummies Technology/General User

Composition Services

Debbie Stailey, Director of Composition Services

Contents at a Glance

Table of Contents

Part V: The Part of Tens 315

Chapter 20: Ten Essential Steps toward a Successful Wind-Electric System 317

Chapter 21: Ten Wind-Energy Mistakes.........................323

Chapter 22: Ten Tales of Wind-Energy Users and "Abusers" 327

Introduction

- -

*W*ind energy is without a doubt the most difficult renewable resource to capture. At the same time, it often ends up being the most attractive. Should you be discouraged by the difficulty of the job? No! But if you don't take the difficulty seriously, you will be disappointed. In my many years of working with wind-electric systems, I've seen many, many problems and failures. Some of these were due to equipment design flaws or freaks of nature, but most were due to poorly designed systems and poor maintenance.

This book doesn't sugarcoat the technology or the industry. This is to your benefit! I suspect that your goal is the same as mine: a *successful* wind energy system that gives you electricity that's cleaner and perhaps less costly than what you're using now. Getting to that goal requires looking seriously at what it takes to generate electricity with the wind for the long haul.

Whether you come to this technology with environmental, financial, independence, or hobby motivations, you can get some or all of your home's electricity from the wind. The wind, driven by natural cycles, originating from the sun, is an abundant and renewable resource. Although you still have the capital and maintenance costs of any energy-generating system, you'll never pay a fuel cost on a wind-electric system. (And a *system* is what you need — not a single component but a wisely designed collection of components that work together to capture wind energy, condition it, perhaps store it, and make it usable to you and your neighbors.)

If you do your homework, find good partners, and design, install, and maintain a robust system, you'll be set up for years of satisfying energy. When the wind blows, you'll smile, knowing that it's working for you.

About This Book

Many people have written books about wind electricity, several of which I use and recommend. This book focuses on a real-world, nontechnical approach to designing and installing wind-electric systems. I didn't write it to turn you into a wind-generator designer, a tower contractor, or an electrician. It's for homeowners who want to explore the possibility of using wind energy and want straight advice from someone with nothing to sell and a great deal of experience with what does and doesn't work.

You don't have to read this book from cover to cover; it's designed so you can dip into and out of any topic at any time. Read what you want, put the book back on your shelf, and bring it down again whenever you need.

If you decide to install your own system, you'll need more than this book. If you decide to hire the job out, you'll be well positioned to ask the right questions, scrutinize the answers you hear, and make wise choices about contractors and system design.

Conventions Used in This Book

To help you navigate this book, I've established the following conventions:

- **Boldface** text emphasizes the key words in bulleted lists and actions to take in numbered lists.
- New terms in this book appear in *italics* and are explained in the text (and often in the glossary in Appendix A).
- All Web addresses appear in `monofont`.

Some Web addresses may break across two lines of text. Where that happens, rest assured that I haven't put in any extra characters (such as hyphens) to indicate the break. When using one of these Web addresses, just type in exactly what you see in this book, pretending that the line break doesn't exist.

What You're Not to Read

Please don't read anything that you think is boring or pushes your buttons. If my writing or opinion or the topic doesn't capture your imagination, move on! Each chapter is written to stand on its own, and there's no requirement to read it all or read in sequence. Also, any text preceded by the Technical Stuff icon or included in a sidebar (a shaded gray box) is extra, and you don't need to read it in order to understand the subject at hand.

Foolish Assumptions

Some wit once said, "Assumption is the mother of all screw-ups." I state upfront my basic assumptions about you so I can help you avoid making a mess. Here's what I assume:

✔ You are interested in successful wind-electric systems.

✔ You want to know whether a wind-electric system is a viable option for your circumstances, and you want a solid grounding in the concepts and components of such a system.

✔ You want your misunderstandings, myths, and fantasies about wind energy to be corrected. You'd rather hear straight talk than sales hype.

✔ Your goals include cleaner, cheaper, or more local electricity.

✔ You know that really valuable things cost — in time, money, and energy. In other words, you know that TANSTAAFL — there ain't no such thing as a free lunch.

How This Book Is Organized

This book is divided into six parts. As a strong advocate of not coloring between the lines, I encourage you to seek out the parts and chapters that you're interested in and read them first. Here's how the general topics are divided.

Part 1: A Wind Primer: Stuff You Need to Know

This part focuses on key wind energy information. It's important to decide upfront whether you're a good candidate for wind energy and to understand the basic parts of the systems and how those parts fit together. Electrical terminology and concept basics can help you understand how these systems work, and understanding some basic wind energy principles can help you understand the resource you're trying to capture.

Part 11: Assessing Your Situation

This part is perhaps the most important part of the book because it takes a hard look at your home and its energy use and your site and its energy resource. Energy efficiency is a vital strategy that helps reduce your energy load and therefore your system size and budget. Understanding wind site assessment helps you get realistic about your site's potential. How you interact with the utility grid — or don't — is covered here, as is economic "payback" and your options if you decide not to use wind energy.

Part III: Assembling Your System

After you've decided to have a wind-electric system, you have a number of choices about system design. An early question is whether you'll be doing this all yourself or working with others. You also need to decide on a wind generator, a tower, and the other components. And then you pull the system design together into a unified whole.

Part IV: Installing and Operating Your System

The culmination of all your design work is the actual installation. Before you start, focusing on safety — with towers, mechanical and electrical aspects, and so on — is step number one. After your installation, you need to learn to live with, maintain, and enjoy your system.

Part V: The Part of Tens

This part, which is a feature of all *For Dummies* books, starts with ten goals for your wind-electric system. These help you get on track. I then outline ten common mistakes so you can steer around them. And ten stories — of successful and not-so-successful systems — give you examples to follow or avoid.

Part VI: Appendixes

In this part, you find a brief glossary full of important wind energy terms as well as abbreviations and conversion tables.

Icons Used in This Book

This book is peppered with the following icons to draw attention to specific concepts:

This icon highlights key theories and practices worth keeping in mind during your design and installation process.

Sometimes I like to show off my technical prowess. You can decide whether to read and indulge me.

Look for text marked with this icon for ways to work or ways to look at things that you may not have thought of.

Safety is the number one priority. Don't ignore the advice you find with this icon.

Where to Go from Here

Scan through the table of contents and see what excites you, and then dive in. If you're determined to be organized, read straight through. But this book isn't entirely linear. As with my teaching style, it's more circular, with recurring themes. This isn't because I forgot I've already said something but because some ideas bear repeating.

If you want a quick overview, read Chapter 1, which summarizes the key concepts in this book. If that's too much, check out Part V, the Part of Tens, where the chapters are bite-sized and pithy. Wherever you start and however far you go, I hope this book will help you become realistic about wind-electric systems. If you follow the advice here, you'll be well positioned to capture an abundant, free, and dynamic resource!

Part I
A Wind Primer: Stuff You Need to Know

The 5th Wave   By Rich Tennant

"Get in the cellar, Ma — twister's heading this way, and it's a big one!"

In this part . . .

This part gets you off to a good start in understanding wind-electric systems. Chapter 1 gives you an overview of wind energy. In Chapter 2, you look at your motivations and goals, common objections and legal issues, and your chances for success. Chapter 3 identifies the components of a typical wind-electric system and how they can be put together. To wrap up, Chapter 4 gives you a foundation in electrical terminology, and Chapter 5 covers wind-energy principles.

Chapter 1

Introducing the World of Home Wind Electricity

*H*ave you ever watched a wind generator spinning in the breeze and wanted one for your home? You're not alone. Wind generators — big and small — are captivating. Something about capturing the elusive, invisible force of wind excites people. I'm here to help you take that excitement and succeed in making meaningful amounts of electricity from the wind for years to come. Think of this chapter as your introduction to the wide world of wind energy.

Figuring Out Whether Wind Energy Is Right for You

People chase after wind energy for a variety of very different reasons. Yours may include reducing your impact on the environment, saving money, increasing the reliability of your home's electricity, boosting your social status, or adopting a fascinating hobby. Being clear about your motivation can help you make sure you reach your goals. For example, if wind energy is a hobby, you'll be less concerned about payback; if it's an environmental passion, you'll want to make sure you're actually cleaning up the Earth, not burdening it with more stuff.

To be successful, you obviously need wind, and you need a good site where you can install a tall tower to get up into the good wind. But you also need to please your local bureaucrats, as well as your family and neighbors. Educating yourself about the common objections to wind energy can help you educate others about the reality of a wind-energy system and get them on your side.

If you have the right site and situation, you still need to have or hire the skills to design, install, and maintain a system. Someone — either you or someone you hire — has to be hands-on. Someone needs to be ready to take on a system that isn't easy and may well give you headaches at times. Finding mentors, experienced wind-energy users, and professionals can help. Chapter 2 explores these issues and more in detail.

Understanding the Components of Wind-Energy Systems

If your goal is to make wind electricity, you need more than a wind generator. A wind-electric system, even at its most basic, includes the following parts:

- ✔ **Wind generator:** The spinning device that captures wind energy and converts it to electricity

- ✔ **Tower:** The steel structure that holds your wind generator up in the wind

- ✔ **Transmission:** Wire and associated equipment

- ✔ **Controls:** The charge controller, inverter, and so on

- ✔ **Batteries and/or grid interface:** Equipment for energy storage and grid interconnection

- ✔ **Metering, disconnects, overcurrent protection, grounding, and more:** Gear to keep track of your system's performance and keep the system safe

Chapter 3 details these system components so you can understand their functions and be ready to think about how to put them together in your system.

Focusing on Electricity Fundamentals

You can't do a good job of designing, installing, and operating an electrical system without understanding electricity. If you don't know a watt from a volt, Chapter 4 helps you with plain-language explanations of electrical terms, including the following:

✔ **Wattage:** The rate of energy generation, transfer, or use

✔ **Watt-hour:** The unit of electrical energy

✔ **Voltage:** Electrical pressure

✔ **Amperage:** The flow rate (often called *current*) of electrons (charges)

✔ **Direct current (DC):** One-way flow of electrons

✔ **Alternating current (AC):** Two-way flow of electrons

✔ **Amp-hour:** Battery storage capacity

✔ **Ohm:** Resistance to the flow of charges

✔ **Hertz:** The frequency of AC alternation

Perusing Wind-Energy Principles

In addition to having a clear understanding of electrical principles, you want to understand wind-energy principles. Wind is invisible and a bit mysterious. Its power increases with the wind speed cubed (V^3), which is a particularly important concept to understand; it means that a small change in wind speed can have a big effect on how much electricity you're generating.

Capturing wind energy requires a large enough collector on your generator, and air density (and therefore elevation) also has an effect on power. (One wind power formula says that the power in the wind equals $1/2$ times air density times the collector area times the wind speed cubed, or $P = 1/2DAV^3$.)

Although this formula helps you understand the comparative power available in the wind, energy (measured in watt-hours) is the prize you seek, so focusing on instantaneous or maximum power (measured in watts) is a distraction at best. And you don't want watt-hours for a day, week, or a year but for years if not decades, so you should seek reliable equipment that churns out the watt-hours for the long haul. Chapter 5 explores these topics and more.

Getting a Grip on Your Energy Situation

Because your goal is undoubtedly to make some or all of the electrical energy for your home, understanding how much energy you use is crucial. The next step is to work on energy efficiency — making the best use of your energy. Your site's wind-energy potential needs careful consideration, too. All these steps improve your chances of generating the amount of energy you need.

Before designing a system, you need to decide how your system will or will not relate to the utility grid. The grid allows you more leeway in how much energy you can use or need to produce, but hooking up does involve getting through some red tape.

After you calculate how much energy your chosen turbine is likely to produce on your site, you can get an idea of the value of your investment. If you decide not to go after wind energy, you have some other options. Read on — this section covers all these energy issues.

Conducting an assessment and increasing your home's efficiency

An energy assessment can give you a big-picture view of the energy use of your family, including specifics about each energy user in your home (by *energy user,* I mean your air conditioner or stereo, not, say, your spouse and kids). The simplest way to get the overall view of your electricity usage is from your utility bill, but getting more specific can help you ferret out the biggest culprits.

Chapter 6 gets down to the nitty-gritty for you. Ideally, you'll end up with a detailed list of all your home's energy users, with an accurate number of kilowatt-hours (kWh) per day attached to each. You'll also figure out which appliances are using energy even when you aren't using them so you can make wise energy-efficiency moves.

If you're motivated to save money or the environment or both, focusing on energy efficiency is your absolute best move. Energy efficiency has a larger return on your investment and a lower upfront cost than any wind-electric system does. Chapter 7 talks you through the details of improving your lighting, appliances, and other energy hogs in your home. The end result may be cutting your energy use by one-third to one-half or more. This means that your wind-electric system — and its budget — can shrink accordingly.

Calculating your home's potential for wind energy

Casual observation of the wind is almost useless as information for your wind-electric system design. Though getting specific numbers may be difficult and costly, you need to work toward getting as accurate an estimate of your average wind resource as possible. Wind is your fuel, and without knowing how much fuel you have, you have no way of knowing how much electricity you can generate.

I describe various subjective and less-subjective methods of estimating your wind resource in Chapter 8. I recommend that you use as many of them as possible and still be conservative with your estimate. Your wind resource is the biggest factor in what you get out of your wind generator, and there's no sense in setting yourself up for disappointment with enthusiastic estimates.

Knowing your home's relationship to the grid

You have a very basic decision to make in designing your system: how the utility grid will be involved. Tying into the grid allows you to buy energy from the utility when the wind isn't blowing and to send energy to the grid when your turbine is making more energy than you can use, usually for an energy credit. You have three basic choices, assuming you aren't already far from the grid:

- **Batteryless grid-tie:** This is the most efficient, effective, and economical system, but when the utility is out of service, your system is, too.
- **On-grid with battery backup:** You have the benefit of using the grid as a "battery" to get credit for your excess electricity, but you also have outage protection.
- **Off-grid:** You have to make all your own electricity all the time.

Chapter 9, which delves into the pros, cons, and configurations of these systems, can help you make your decision.

Determining payback on your investment

Everyone wants his or her purchases to make sense. Financial sense is one way to look at what you buy, though actually it's not the most common way. Almost all your purchases include other values — quality, style, the environment, personal tastes, and so on.

Chapter 10 looks at both the financial side and the full value of purchasing, installing, and using a wind-electric system. I challenge you to look hard at your motivations, your decisions, and the alternatives to renewable energy.

Looking at other energy options

Wind energy is not for everyone, and it may not be for you. Consider these circumstances:

✔ Sometimes you just don't have the right site. You need enough wind, you need legal permission to put up a tall tower, and you need the physical space to do it.

✔ Sometimes you don't have the right situation. Electricity from the utility may be very cheap in your area, or incentives (such as subsidies, loans, and tax breaks) may be very low or nonexistent.

✔ Sometimes you're not the right type of hands-on person, or you don't have the dough to hire the right type of person to keep a system going.

In those cases, you don't need to give up on your renewable energy dreams. Chapter 11 presents some other options:

✔ **Solar electricity:** Photovoltaic (PV) modules — solar-electric cells — provide clean, reliable electricity for decades.

✔ **Hydro electricity:** If you have a stream falling down your back forty, you can tap it for electricity.

✔ **Solar thermal applications:** Heating your home and/or water with sunshine are definite options.

✔ **"Green power" purchases:** Your utility or other providers may offer greentags, renewable energy certificates (RECs), or other ways to use your dollars to support clean electricity projects.

✔ **Transportation alternatives:** From bicycles to hybrid cars, transportation is a field ripe for energy savings.

✔ **Simplifying your life:** Tackling this toughest of jobs may reap the largest reward at the least cost.

Designing Your Wind-Energy System

System design pulls together the information from your energy assessment, site assessment, and personal assessment. With this information, you need to find suppliers, contractors, and other team members. Then you actually need to choose the components you'll use — the wind generator, tower, and balance of systems (BOS) gear. Add the team to the gear and work through the process of figuring out how a gang of people and a pile of gear can turn into a working system.

A team of experts to help

Chapter 12 can help you decide where you'll be along the continuum of sole owner-installer to check-writer and observer. In all cases, you'll be dealing with other people; even if you're going it alone on the design and installation, you still need to buy your equipment from someone.

I recommend buying from someone who can supply you not only with gear but also with information about how to design, install, and maintain a long-lasting system. Paying a bit more for the gear is worth it if you end up with a consultant you can turn to when you have questions.

If you're new to this field, you're more likely to be somewhere on the other end of the continuum, where you'll be either working as part of a team or hiring a contractor to do the whole job. Picking your team carefully can mean the difference between a delightful experience and a disaster.

Wind generators

In a wind-electric system, the wind generator is the star of the show. Of course, there are no one-man or one-woman shows in wind energy — you need all the components. So I'd encourage you *not* to turn to Chapter 13 first and drool over your wind generator options. Get a good grounding in the whole system first so that when it's time to look at the star, you understand the whole script and performance.

Being clear about what size of wind generator you need is a good first step. This goes back to your energy *load* — how much energy you use, because the wind generator's diameter will be directly related to how much energy you want to generate. Then looking at the different turbine configurations is worth doing (although most configurations will be similar, and the oddball ones are best left to the crazy experimenters).

Before you make your final pick, look at what kind of an owner you'll be, how harsh your site is, and what your budget constraints are. I discuss all this and more in Chapter 13.

Towers

Wind-generator towers are the most ignored and underrated part of wind-electric systems, but they're crucial for good performance. Because the power available in the wind increases with the cube of the wind speed (V^3), getting just a bit more wind means getting a lot more energy. And the way to get a bit more wind is to put up a taller tower. The standard rule of thumb in the industry is to have the lowest blade *at least* 30 feet above anything within 500 feet.

Your site will help decide what tower style you need, based partly on the *footprint* (the area taken up by tower, anchor, and guy wires). Towers come in three basic styles, with variations on the basic three. Table 1-1 shows how the styles compare.

Table 1-1	Basic Tower Styles		
Tower Type	*Need to Climb*	*Footprint Size*	*Cost*
Tilt-up	No; all maintenance is on the ground	Largest footprint	Medium cost
Fixed guyed	Yes	Medium footprint	Lowest cost
Freestanding	Yes	Smallest footprint	Highest cost

I discourage you from using homebuilt towers or mounting a turbine on your roof. If you want reliable, safe performance, your best bet is to go with the tried and true, unless you're a crazy inventor who loves to experiment and is willing to take the consequences. Also keep in mind that choosing a tower that's too short and too close to obstructions (such as trees, buildings, and landforms) is the number one mistake that people make in small wind energy today.

Choosing your tower type, height, and overall plan is vital to achieving good results. Your tower is a big chunk of the budget, and tower installation takes a significant amount of time. Turn to Chapter 14 to delve into towers.

Other components

The supporting cast in a wind-electric system may include a charge controller, inverter, disconnects and overcurrent protection, wiring, batteries, and more. Together, these items are called *balance of systems* (BOS) equipment. There's a lot of detail to understand here, and unless you already have experience under your belt, you need support in this realm. Chapter 15 can help you know which questions to ask and how to understand the answers.

Tying everything together

Understanding what each component does is one thing. Designing an integrated system of components that work together well to accomplish your goals is another thing altogether.

Here are some specific decisions about your system and the factors that impact them:

   ✔ **Wind generator size:** Your energy load (kilowatt-hours per day/month/ year) and your wind resource (average wind speed) determine the size of your wind generator.

✔ **Tower height:** Your site and its wind shear help you determine the minimum tower height — I recommend going higher than the minimum.

✔ **Relationship to the grid:** Your situation, motivation, and mindset determine your system type. Choose wisely among off-grid, battery-based grid tie, and batteryless grid-tie. Changing your mind later often isn't easy.

Choose your project team carefully, too. Don't fall victim to the pain of the "best price," which is often attached to the worst service, if not the worst product. Consider the advantages of buying a complete package from a dealer who can support you. I cover these issues and others in Chapter 16.

Installing and Using Your Wind-Energy System

When it's time to get your hands dirty — or someone else's hands dirty — and install your wind-electric system, the first priority should be safety. Wind-energy systems have more dangers than other renewable energy systems and more than a typical home's electrical system.

Installation needs to be done with care, and it requires experience in a variety of fields, from concrete to mechanical construction to electricity. After that, monitoring and maintaining a system can take less technical savvy if you've done a good job of design and installation.

Staying safe

Hazards on wind-energy jobs include gravity, the weather, mechanical moving parts and failure, electricity, batteries, and live (human and animal) hazards. To work safely around these hazards, you need a variety of safety equipment. Tower climbing has its own set of tools that includes harnesses, lanyards, fall-arrest equipment, tool bags, pulleys, lines, and more.

Knowing how to work safely on the tower and on the ground is not optional. You're dealing with life and death here. Falling or dropping something from even 40 or 50 feet can be fatal. Understanding how to work safely around electricity, batteries, and mechanical tools and components is also important.

Chapter 17 gives you a good start on understanding the dangers and the gear, and it sets you up to find out more on the job.

Installing your system

Wind-electric system installation covers a variety of trades:

- ✔ You start by laying out the tower, excavating, and pouring concrete for the tower base and anchors.

- ✔ Tower installation includes assembling and lifting, with more than one possible approach.

- ✔ Wind-generator installation also involves weight, mechanical work, and electrical connections.

- ✔ Electrical work spans the distance from the tower top (where you find a whole different set of working conditions) to the power room.

Chapter 18 gives you the big picture on these processes, which vary depending on your specific site, design, and equipment.

Monitoring and maintaining your system

Living with a wind-electric system is a satisfying occupation. A wind generator gives you an eye on the sky and the weather. Your awareness should extend to the instantaneous output (watts) and cumulated output (watt-hours) of your system. Not only will your metering and data-logging systems be fascinating to keep track of, but they'll also let you know how your system is performing.

Wind energy is not a build-it-and-walk-away proposition. These systems require regular maintenance. At least once a year — and maybe twice — someone needs to lower or climb the tower to check the wind generator, do all scheduled maintenance, and address any problems. On the ground, batteries are the primary focus for maintenance, but you need to be aware of other system components and their calibration, too.

Chapter 19 covers these issues and more. If you're aware of your system and treat it carefully, you'll be set up for a long career capturing the energy in the wind.

Chapter 2

Is Wind Energy for You?

Making electricity using the wind can be extremely satisfying, or it can be very frustrating. It can be environmentally friendly or high-impact. And it can be cost-effective or a dollar sink.

People come to wind energy with many motivations and goals. Others come to wind energy with objections. What do you really need to be a successful home wind farmer? First of all, you need a reasonable wind resource and a good site. Then you need the appropriate attitude, education, and experience to tackle buying, perhaps installing, operating, and maintaining a system. This chapter helps you decide whether wind energy is a good match for you.

Exploring Motivations for Using Wind Energy

People have many different motivations for wanting to use wind energy. Perhaps they're concerned about the environmental effects of their current fossil-fuel systems, or maybe these systems have become unreliable. They may want to try wind energy to save money or to be on the forefront of the renewable energy movement because they're fascinated with the technology.

Take a few minutes to determine why you want to use wind energy before you spend your hard-earned cash and precious time on a system. In this section, I describe the most common motivations people have for using wind energy.

Green reasons: Living more sustainably

Many people come to wind energy — and to renewable energy in general — with an environmental motivation. North Americans are gradually discovering that the last 200 years have been an anomaly. They've been blessed (or cursed) with cheap fossil-fuel energy, and in the not-too-distant future, they'll gradually have to return to the energy cultures used for thousands of years — energy from sun, wind, water, and *biomass* (plant and animal products).

People are getting tired of this supposedly "cheap" energy they rely on so heavily today. When you look at all the true costs and impacts, it's not cheap at all. For example, a large percentage of electricity in the United States is made with coal, which results in removal of mountaintops in Appalachia and strip mining of large tracts of open land in the West. How cheap is this really?

You don't need to be a rocket scientist to see that renewables (including wind energy) have two distinct advantages over nonrenewables:

- ✔ **Long-term costs:** Both technologies have capital costs and maintenance costs, but renewable energy systems bypass one of the largest ongoing expenses of nonrenewables: fuel costs.

- ✔ **Environmental impact:** Although all technologies and people have an impact, renewables' impact is modest compared to that of nonrenewables. Coal mining leaves behind damaged land and decapitated mountains, as well as polluted air. Nuclear energy leaves dangerous waste that needs containment for hundreds of years. Gas and oil generation result in contaminated land, air, and water.

    On the other hand, the impact of renewables, though not zero, is minor. These systems have *embodied energy;* that is, you need materials and processes to make them. And they can be built without regard for environmental damage. But even if constructed in a dirty way (and most aren't), they have little ongoing impact on the environment after installation. Wind is not damaged or depleted by wind generators, and the surrounding environment sees only minor (and subjective) impact.

Individuals who value clean energy — not just cheap energy — put short-term cost aside and continue to be willing to pay more and exert more effort to get it. Just as people value organic food, nontoxic finishes, and such, they value energy that has a low environmental impact.

Getting a backup power system

Most city dwellers in North America take the reliability of the utility grid for granted. Outages are actually quite rare, and when they happen, they're generally of short duration. But where the grid is less reliable in rural environments, a large minority of people do need backup. And many people want backup even if they don't really need it.

When I need to find out whether a wind-electric system client is looking for a backup system for utility outages, I ask three questions:

- ✔ **How often do you have utility outages, and how long do they last?** In many urban and suburban settings, the answer to this question tells me whether the grid is reliable. Personally, I'd avoid battery backup unless I had a dozen-plus outages a year, but you may be happy to pay for a backup system to protect against only a few outages.

- ✔ **Do you have backup loads that are critical to life, health, and safety?** This question can identify folks on medical support such as dialysis or oxygen who may require very reliable, 24-hour electricity.

- ✔ **How do you react during an outage?** If outages are significant, this question tells me whether the person becomes alarmed because of the lack of electricity or pulls out candles and enjoys a relaxing ambiance.

With batteries, wind-electric systems can provide stable, 24/7/365 electricity. These systems can produce electricity of higher quality than the utility grid and with higher reliability. Batteries mean backup systems are higher cost and take maintenance, but if reliability is high on your list of motivations, they can satisfy. Asking yourself the preceding questions can help you decide what system configuration you need (see Chapter 9 for more on this).

Saving money

Many people approach wind electricity in the hope of saving money. In fact, I'd say that most people are unrealistically hopeful about the financial benefit of small wind-electric systems. Some even think they'll be getting a big check every month, which is unrealistic for home-scale turbines. On the utility-scale, wind farms are reliable money makers; on the home-scale, things aren't so rosy. In between, there are some possibilities for financial return as well.

But everyone wants to save money, right? So, rather than tossing out this motivation, you may just need to temper it and take a long view. Energy efficiency will almost always be a better investment than wind energy, but if the right conditions collide, you may see a 3 to 10 percent return on your investment if you install and maintain well. In rare cases with excellent wind resources and incentives, it may be better. And if line extension costs are high to your off-grid property, a wind-electric system can indeed save you a lot. However, home wind-electric systems aren't primarily moneymakers.

In many cases, you have to look at the large picture of "true value" rather than simply dollars. How much money you'll save depends on your wind resource, your tower height, the size of your turbine, your utility rate, and how well you design and install your system. Check out Chapter 10 for details on calculating your payback on a wind system.

Experiencing the fun of doing it yourself

Some people go for wind electricity just because they like playing with the technology. I call these folks "wind-electric gear heads." They love putting things together and keeping them running.

One strain of this do-it-yourself virus is the homebrew disease. All over the world, folks are finding out how to build their own wind generators from scratch, often with salvaged parts. Many follow plans or books from the likes of wind energy experts Hugh Piggott of www.scoraigwind.com or the otherpower.com Dans (Dan Bartmann and Dan Fink — see Chapter 22 for their stories). Others make it up as they go along. Results vary, of course, but it's quite possible to make a durable, functional, productive machine yourself.

Even if you buy a manufactured wind generator, wind-electric systems can bring out the hobbyist in you. The whole project lends itself to hands-on folks who are ideal for home-scale wind, because the technology isn't mature enough to be trouble-free.

Being on the cutting edge

A distant cousin of the do-it-yourselfer, the cutting-edge class of wind-energy users wants to be on the forefront of technology. The attraction is not for environmental reasons, reliability, or cost but because wind energy is the latest thing. This sort of motivation turns up frequently in other fields — cars, computers, clothing. So why not energy technology?

Your own system can offer you several opportunities to keep up with the advances in the industry. The turbines themselves are becoming more sophisticated in design, construction, and application. And the related electronics — including those for power conditioning and monitoring — give ample opportunity for nerding. Don't underestimate the this-is-cool factor when looking at your motivations.

Increasing your self-reliance

America was built by independent people who wanted the freedom to live their lives as they pleased. Similar leanings toward self-reliance lead people to install wind generators. Even on the grid, generating some or all of your electricity with the wind makes you less dependent on outside entities. If you add batteries to a system, you can provide backup when the utility is down or go completely off-grid — the ultimate in self-reliance.

Balance your desire for independence against the increased cost of systems that have batteries and may not have the grid. My advice is to combine your desire to take care of yourself with a cooperative relationship with the grid to get the best of both worlds.

Wind-energy user lifestyles

People from all walks of life use wind electricity — no social or economic profile required. I try to avoid assumptions about people who want to tap into wind energy. For one thing, I often can't accurately guess what peoples' motivations are. For another, if I lean toward a different motivation, I may offend the interested people, discouraging them from using wind energy. And in the end, my goal as an author, teacher, and advocate of renewable energy is to promote renewables, not a particular motivation for renewables. Still, I think that most wind-energy users fit into one of three general lifestyles:

- ✔ **Off-grid hippies and yuppies:** Many people want to get away from it all and live in the woods. These can be scruffy back-to-the-landers like me, who bought a piece of forest in the early 1980s and gradually built a home for my family. But it can also include up-and-coming professionals, retirees, and everyone in between — folks who share a desire for space, quiet, and nature.

  Home wind systems can be tiny, for the energy sippers, or very large, for the folks who want to live in the woods with a city lifestyle and energy appetite. Wind-electric systems can run a few lights and a cassette deck to play the old Grateful Dead tapes. They can also run full-featured remote homes with all the trimmings.

- ✔ **Farmers and ranchers:** Farmers and ranchers are perfect candidates for wind-electric systems. These people already know how to keep the tractor, truck, baler, and other farm gear running, so they have the proper mindset for wind-energy systems. They also have the ideal topography — wide open spaces of agricultural or grazing land, with fewer obstructions that slow the wind. They also typically have fewer legal, aesthetic, and neighbor obstructions to deal with, because the authorities and communities are already used to semi-industrial equipment and lifestyle.

- ✔ **City and suburb dwellers:** This is the toughest setting for wind energy, because obstructions are at the maximum and space at a minimum. Capturing wind energy here isn't impossible, but it's very difficult. People in the urbs and the suburbs should proceed with great caution and focus on high-quality resource assessment. Few suburban sites and very few urban sites yield highly productive wind-electric systems. I long for a day when homeowners in these settings can effectively tap wind energy because tall towers become accepted by neighbors and bureaucrats as the norm, just as utility lines and propane tanks are accepted. Without tall towers, they are doomed to disappointing results in most cases.

Meeting Some Minimum Requirements for a Wind System

Is wind electricity for everyone? No! In fact, most people don't have the right situation to tap the energy in the wind. Before you start opening your wallet, assess your situation. In this section, I talk about the prerequisites for a successful wind-electric system. (And if you decide that wind power isn't for you, check out Chapter 11 for some other green energy options.)

A wind resource

This fact seems so obvious that even a politician would see it, but many higher and freer human life forms overlook it: If you want to make wind electricity, you need wind. And if you want a significant amount of electricity, you need a significant amount of wind.

What's enough wind? Depending on the application and situation, you want an average wind speed between 8 and 14 miles per hour. (Read Chapter 8 for details on this subject.) Make no mistake — you need quite a windy site to make financial sense with a wind-electric system. And that always means a tall tower, well above local obstructions.

Wide, open spaces

A tall tower means you need a suitable site for installing the tower. This isn't going to happen if you live in a row house in the Bronx. It's also not going to happen if your local bureaucrats restrict tower heights to 40 feet. The standard industry rule of thumb is to install a wind generator so the lowest blade tip is *at least* 30 feet above anything within 500 feet. Higher is always better. This means that in most cases, towers are in the 100- to 200-foot range because of houses, trees, and other normal obstructions.

You need space to accommodate the footprint of the tower (see Chapter 14 for details). The bottom line here is that you'll be hard pressed to install a wind turbine on less than about one acre in most circumstances, often for bureaucratic reasons and sometimes for cost.

The human factor: Skills and support

Succeeding as a home wind farmer isn't easy. Wind-electric systems are more difficult to maintain than, say, gardens. You have a higher probability of harvesting good carrots than of harvesting wind energy. When determining your overall chances of wind-farming success, you need to consider your skills, your outlook toward maintenance, and the people you can find (or hire) for help. The following three qualifications should be high on your list.

Are you hands-on?

Are you the hands-on, do-it-yourself guy or gal I talk about earlier in this chapter? Or is your wallet fat enough to pay a professional to maintain and repair your system for its full lifetime? If so, you have the qualifications to consider a wind-electric system.

Home-scale wind-electric systems are not simple, easy, or even terribly reliable when you look at systems in the field today. In my 25+ years rubbing elbows with wind turbines, I have not seen *one* system that has been trouble-free for its lifetime. Problems have varied from irritating to catastrophic, but in all cases, there have been problems. You should assume that this will be the case with your system — there are no maintenance-free wind turbines. Chapter 19 delves into common operation and maintenance duties.

If you're not a hands-on hobbyist, you need to either become one or hire one. Small wind-energy systems are not of the install-and-walk-away variety. I'm not sure they ever will be, but in a relatively young industry, you shouldn't expect this. Flip to Chapter 12 for help on finding knowledgeable experts.

Are you ready to be a beta tester?

A young, immature industry produces immature products. There are so few turbine manufacturers and so few turbines manufactured that working the bugs out of a product takes some time.

It's normal for some of the product beta testing to be done on the backs of the purchasers. This is the unfortunate truth with most home-scale turbines today. As much as the better manufacturers try to test before the release of a product, small companies with limited cash flow need to get products to market in order to stay in business. Plan on it: You will be doing some beta testing for someone. (You can limit your risk, though, by avoiding unconventional products or designs; see Chapter 21 for details.)

Do you have a mentor?

Few people reading this book would install their own furnace or swimming pool systems. And yet too often, people think that wind-electric systems should be easy to install — color-by-number and follow the directions.

Wind-electric systems are more complicated (and dangerous!) than furnaces or pools. Some aspects of design and installation require years of training and experience. A wind-electric system is not a gas grill with step-by-step directions; rather, it involves a complicated design and construction process.

Don't — I repeat, don't — try to design or install your system alone the first time, even if you are hands-on. Find yourself a mentor, partner, contractor, or experienced friend to help. You need to buy your equipment somewhere. I recommend buying it from a company that can help you all the way through the design and installation process. Even if you do most or all of the work yourself, having an experienced mentor on tap is a great help in putting together a safe and effective system. Parts III and IV walk you through the process of safely assembling and installing a wind system.

Dealing with Common Objections

Wind energy has had some bad press over the last few decades. Some of it's deserved, but most is the result of misinformation, hype, or competing interests that want to discredit wind energy. Too often, I find that people are looking for problems where there aren't many, or they're trying to find excuses not to pursue harvesting renewable energy. In this section, I set the record straight on typical objections to wind-energy systems.

Small capacity and inconsistent power

I frequently hear newbies to wind energy object that they need electricity even when the wind isn't blowing and that wind turbines can't run the big appliances in their homes. Are these valid concerns?

All renewable energy is variable, and wind is perhaps the most variable. This doesn't have to be a problem, though it does have to be a consideration. The strength of this objection depends on what sort of system you have and which backup measures you use (see Chapter 9 for details):

✔ If you're off-grid, you want some electricity 24/7 and electricity on demand during much of the day. If you were to install a wind generator and wire it directly to your house's electrical loads, you'd be pretty unhappy — imagine your lights going off or your computer crashing every time the wind died down. But off-grid systems have batteries and backup generators, so exactly when the wind is strong, weak, or nonexistent isn't terribly relevant. In a well-designed system, the batteries cover the baseline loads when wind isn't sufficient, and another generating source or a backup generator picks up the slack.

✔ In on-grid systems, the utility company's grid acts like both a battery and a generator — accepting your system's excess energy and providing electricity when you need it — so the wind's variability is not an issue.

But what about capacity? Is wind capable of running everything in your home and shop? Sure. Wind electricity could run everything in the world if people installed enough generating capacity. Can it cover the peaks of your energy use? Not all the time, but this is part of why systems have batteries and/or are connected to the utility grid. Don't expect a tiny wind generator to run a huge houseful of energy suckers, but certainly if you invest in a large enough machine, you can power homes, businesses, and industries with wind. (Of course, your best bet is to make your home more energy-efficient before you start shopping for turbines — see Chapter 4 for details.)

Noise

Are wind turbines loud? That's kind of like asking whether cars are fast. Your sister-in-law's convertible may be able to peg the speedometer, whereas that broken-down jalopy in your back alley can't keep up with the freeway traffic. My point is that you should be careful about generalizing about wind turbines (or anything). Some wind turbines are loud, and some aren't.

Poorly designed machines can be extremely noisy. Some microturbines have been compared to chainsaws. Others are so well-designed that you have to tune your ear so you can hear them over the ambient noise.

Note that in some specific conditions, more sound will come from a turbine. In higher winds, turbines *govern* or *furl* (move blades out of their best aerodynamic position) to spill excess energy, and at times, this can be louder than normal operation.

One primary factor has the greatest influence on turbine sound: rotational speed. When comparing turbines of the same general size, look up the rotations per minute (rpm), and you'll find in general that the lower rpm machines are quieter. Smaller diameter machines are often designed to run at higher speed, so they tend to be louder.

Sound is subjective. My advice is to actually *hear* the turbines you're considering buying before you make a decision. Visit an operating example of each turbine on a windy day. Come back on a day with light winds. Talk with the owner and neighbors. Find a machine that you and your neighbors can live with. It's possible; you just need to do your homework and shop intelligently. Chapter 13 has plenty of information on turbines to get you started.

Design and installation dangers

Wind-electric systems can be dangerous. These systems offer more hazards than any other type of renewable energy system. Your renewable energy system won't look quite so "clean" if it injures or kills you or someone else.

Are these dangers insurmountable? No. But if you treat these systems as if the hazards don't exist, you won't be able to guard against them. Keep the risks in mind as you design and build your system. If you approach wind-energy systems with safety as your highest priority, you can live near them and enjoy them. See Chapter 17 for details on the dangers and how to avoid them.

Potential problems with birds

I remember Tom Gray of the American Wind Energy Association (AWEA) saying that it's easy to compare wind farms unfavorably to *nothing*, because *nothing* has no impact. This is certainly true of the issue of birds and wind turbines.

Few people want to kill birds indiscriminately, so the thought of wind turbines killing birds is something they don't appreciate. But I'll try to bring some reality to this discussion. First, how many birds do wind turbines kill? Studies show that the huge utility turbines kill only one or two birds per year per turbine. These are not bird blenders, and birds are not blind idiots. Home-scale turbines are a tiny fraction of the size of utility-scale turbines, and avian mortality scientists conclude that the issue of birds and home wind turbines isn't even worth studying because it's so insignificant.

More-important questions to ask are the following:

- What is the wind energy replacing?
- How does competing energy technology affect birds?

In other words, how many birds are killed by coal and uranium mining, oil exploration, fossil-fuel burning, and habitat destruction? When you compare apples to apples, wind electricity looks less like a bird killer and more like a bird saver. This is why the Audubon Society endorses responsible wind energy (www.audubon.org/campaign/windPowerQA.html).

Gaining some perspective on how other human activities impact birds is worthwhile. Various studies show that birds killed by automobiles, windows, and housecats make even utility-scale wind turbine bird kills look like a drop in the bucket.

Aesthetics

Wind turbines are ugly! Have you heard that before? I let it roll off my back just like comments about my personal clothing style (which runs to jeans and flannel). Aesthetics are very personal, or as a Latin maxim says, *de gustibus non disputandem est* ("there's no accounting for taste").

Personally, I find wind turbines to be beautiful, and it's not just skin deep. It's great to look at something that's vivacious and powerful while it creates clean electricity. If you want to see ugly, look at utility poles and lines, substations and high-tension lines, propane tanks, refineries, and strip mines. I'm underwhelmed by the number of reports I hear from people who want these in their backyards.

In the end, aesthetics may play a part in your system design process. I encourage you to put together a system that isn't offensive to your aesthetic senses or those of your friends, family, and neighbors. Part III explains how to assemble the system that best meets all your needs.

Lower property values

As wind-electric systems become more common, people are becoming more realistic about their value. These systems used to be considered funky and homebrew and a detriment to property values. But more and more, homeowners, home appraisers, banks, and taxing agencies realize that a home with a built-in electricity supply is of higher value. In fact, many people are beginning to see living in a renewable-energy neighborhood as a sign of high status.

Getting the Green Light

Wind electricity fanatics seek freedom from utility bills, independence from the grid, and the soaring feeling of making their own energy. Too often, they're brought back down to Earth by hungry bureaucrats carving out their turf in the building industry landscape. Figuring out how to surf this bureaucracy may be as challenging as any other aspect of the job.

If you're grid-tied, you also need to deal with the utility, which is a large bureaucracy in itself. As far as government agencies go, the volume and complexity of your entanglement depends on where you live. Some areas require no permits, and others want to see you jumping through many hoops. In this section, I talk about some of the required approvals and restrictions involved in setting up a wind-electric system, as well as some advice on keeping the peace with your neighbors.

Talking to your neighbors first

More important to me than the legal considerations are the social considerations. I want wind energy to be accepted as normal in communities, which takes acceptance by a large slice of the population. I hope you befriend your neighbors and are a polite and friendly advocate for wind energy.

Neighbors can be a thorn in your side — personally and legally. Laws and their interpretation vary widely, but your best bet is to deal with neighbors' concerns in a friendly and personal way to avoid ending up in conflict.

Good neighbors make good neighbors — fences or not. An early move when you're contemplating a wind-electric system should be to *talk* to your neighbors. Show them systems in your region. Let them see and hear turbines on tall towers in action. Show them pictures of the equipment you'd like to buy or even videos of it in action. Talk up your dreams, goals, and plans. Get them on board. This will make your life easier all the way down the line.

One wind energy user I know in Illinois shares the benefits of his system. When there's a utility outage, orange extension cords snake their way to a couple of neighbors so they can power critical loads during the outages. What an ambassador for wind energy he is!

Facing some legal limits

You'll likely have to deal with government employees wanting to control what you do on your own property. You may not like it, but they have the guns, and you have to deal with it. If you, like me, don't deal well with bureaucracy, your best bet may be to hire out dealing with the legal tangle to someone who can do it without high blood pressure. In this section, I discuss some of the red tape.

Zoning requirements

City, county, and state zoning can be very restrictive for wind generators, especially in urban and suburban settings. In some European markets, seeing wind-generator towers right in town, poking well above the houses and trees, is very common. Too often in the States, restrictions drive tower heights down to unreasonably (and unproductively) low levels or prohibit them altogether.

There are places where wind-electric systems just do not make that much sense — densely populated urban and suburban areas are not a great choice (see Chapters 8 and 14 to get more of a sense of whether you have a viable site). However, restrictions often stem from ignorance and fear about the realities of small wind systems. Your local, county, and state governments may all have requirements for you, so you have to research what you're up against.

If you find that your local regulations are too strict, fight them! Capitulating to bad laws produces more bad laws. Instead, use the *good* examples of zoning laws to help change the bad examples. Find allies in your area, in the wind industry, and wherever you can, and band together to improve the regulatory climate for small wind-electric systems. Work together to seek variances and changes in the laws, and set good precedents for tall towers and proper siting of these systems.

Wind politics and the law

The point of installing a wind generator is to make electricity, and more and more governments are getting on the green bandwagon. When they at the same time restrict tower heights, they're taking the greenness down a notch. Philosophically, I prefer to let people experiment, win or lose on their own, and take responsibility. But if governments must have restrictions on tower height, they should be on towers that are too short to be effective and will end up disappointing the owners. I'd suggest restricting systems that perform poorly, like roof mounts and 33-foot towers.

I have a radical idea that wind generator towers shouldn't be set to a higher standard than less-renewable utilities, so I hope you'll fight set-back requirements, too. People live with utility poles and lines very close to their homes, roads,

businesses, and so on. I know of no evidence to suggest that wind generator towers are more dangerous, and yet it's very common to see governments requiring them to be set back from the property lines 1.0, 1.2, or even 1.5 times the tower height.

Laws should protect the rights of citizens to make renewable energy on their properties in the most cost-effective way, and that generally means tall towers set close to dwellings. Do towers ever fall? Certainly — all industries have accidents. But that's why you have tower engineering and homeowners' insurance. Your wind generator tower won't be any more dangerous than your car, and you have permission to drive it about without foundation or guy wires and to go at very high speeds.

Building permits

Extremely rural areas may have no building permit requirements. Cities may have city, county, and state building requirements, though typically you need one building permit from the most local jurisdiction.

The building permit is usually primarily for the tower, which may require engineering for the foundation and the tower itself. Every jurisdiction has different requirements, and you or your contractor has to get a liberal education in how to adhere to them.

Building permits are typically issued by city or county governments. If you have any doubt about where to find these agencies, ask a local builder — builders deal with government agencies all the time and can get you started down the serpentine path.

Electrical permits

State, county, or city authorities require you to get an electrical inspection in most cases. Most jurisdictions use some version of the National Electrical Code (NEC) as their bible. Having someone on your team who understands the jargon and concepts can save you from legal tangles (see Chapter 12 for details on gathering a team of experts).

Qualifying for incentives

Incentives — financial benefits for installing renewable energy systems — vary for every state, area, and region. Incentives may include

✔ Tax breaks — federal, state, or city

✔ Payments for a portion of installation system cost

✔ Payments per kilowatt-hour

Some incentives come from the federal government, some come from state and local governments, some come from utilities, and some come from private organizations. The best resource is the Database of State Incentives for Renewables and Efficiency (DSIRE; www.dsireusa.org), which has detailed information on all renewable energy incentives in the United States.

In some states, homeowners can pull permits and do electrical work on their own homes. In other states, only licensed electricians can do the work. Talk with your electrical authorities or to people in the wind-electric industry in your area to find out what you're facing.

Working with the utility company

If you want to plug your wind-electric system into the utility company's electricity grid, you'll have some interaction with the local utility. (See Chapter 9 for details on the options.) With on-grid systems, utilities range from very supportive to a large pain in the rear end. Some encourage renewables like wind energy and make connecting easy. Others seem to want to make things difficult for all energy producers other than themselves.

Utilities may require you to fill out pages of forms, and their requirements may not be easy for you to understand. When connecting to the utility, your best bet is to find others in your area who have plowed this ground and recruit someone to help you wend your way through the paperwork. And make sure you find the *right people* at the utility. Frequently, there's a designated manager for renewable energy systems, and this person will have clearer answers and higher authority to help than the first person who answers the phone.

Chapter 3

Deconstructing Wind-Energy Systems

. . . .

In This Chapter

▶ Understanding the basic components of a wind-energy system

▶ Looking at the layout of different system configurations

. . . .

*W*hen you look at your neighbor's wind-electric system from the nearby road, it looks like a functioning whole. But when you get closer, you of course discover that it, like any other system, is made up of a variety of pieces.

In any system, having *all* the pieces is important. And you need to understand each of the pieces and make sure they play well together. This chapter gives you the big picture of wind-electric system components and the ways they can be designed into systems that serve various needs. Flip to Chapters 13 through 15 for details on selecting each component for your system.

Before You Begin: Understanding the Importance of Buying a Whole System

When you buy a car, it comes with seats, a steering wheel, tires, an exhaust pipe, and all the other components to make a working transportation system. You'd be disappointed if you found out that these crucial components were optional at an extra charge, like leather seats or the DVD player in back.

With a wind-electric system, getting the whole package isn't quite as easy. Although you can buy a package that has most of the major components, it's rare to find a supplier who sells a complete package with every nut, bolt, breaker, and grounding screw. More often than not, you — or your contractor — need to pull together a package of components, or at least supplement a dealer's package with the odds and ends you need to complete the system.

The pressures of a small and competitive business lead some turbine manu-facturers to promote a product without being clear about all the components you need to use it. Salespeople want to sell, and marketing departments want to market. So don't be surprised if you see a wind turbine advertised on its own for $1,499. When you see the advertising copy telling you the price of the turbine, don't stop there. Ask the hard questions:

- ✔ How much will my tower cost?
- ✔ How much will my electronics cost?
- ✔ How much will my battery bank cost (if you want one)?
- ✔ How much will the total system cost, installed?

When you have good answers to these questions, you're in a position to make a smart buying decision, and you won't end up with a whiz-bang wind turbine in your garage in the box, doing you no good at all.

The turbine is an important player in the system — it's the energy generator. But without all the other components, it's worthless, unless you're looking for some sort of bragging rights. The turbine may end up being only 10 to 20 percent of a complete system's cost. Frequently, the tower, battery bank, and inverter each may cost more than the turbine. The moral of the story? Shop for a system. Buy a system. Install a system.

Surveying System Components

Understanding each system component is necessary so you know how to design and install them to work well together. With each piece of the system, you have choices that affect the overall design and performance of your system. In this section, I run through these components.

The turbine

Your wind generator, or *turbine,* is the wind-energy collector and converter. Its fuel is wind, and its product is electricity. It needs all the other supporting components to operate, and when it has them, it can do its work.

Choosing your wind generator is a very important aspect of your system design. When you know your energy needs (see Chapter 6) and your wind resource (see Chapter 8), you can start to look at turbine options. In Chapter 13, I give you guidelines for this process. In this section, I describe the most important parts of a turbine (see Figure 3-1).

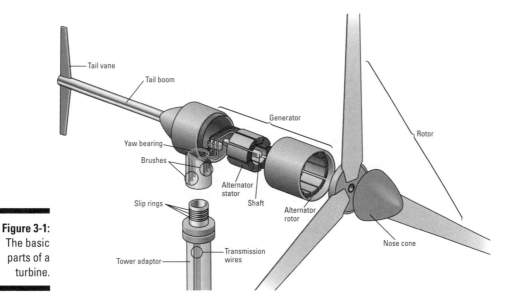

Figure 3-1:
The basic
parts of a
turbine.

The generator

The generator is what makes your electricity. *Generators* make electricity by moving magnetism past copper coils. Most turbines use permanent magnet generators, though some use *wound-field generators,* which create magnetic fields with the rotation of the machine. Each type presents advantages and disadvantages, but these differences aren't huge, nor do I encourage you to base your buying decision based on them.

When choosing a generator, look for simplicity and lack of maintenance, and downplay super-duper efficiency — a bit more blade area can overcome the variations in generator efficiency. Durability always trumps increases in instantaneous performance.

The blades

Your blades and the hub that they attach to make up the *rotor* of your wind generator. Another rotor is inside the generator, but no need to go there: When you hear people talking about a wind generator's rotor, they're almost always talking about the blades and hub.

Most commonly, you see three blades made of wood, fiberglass, carbon fiber, or other high-tech materials. Blades come in many shapes, designed to optimize performance, durability, or ease of manufacture.

The nerds can argue for days about this airfoil or that, but I prefer to let wind-generator designers design wind generators and to head straight for the bottom line: kilowatt-hours over the years. The amount of area swept by the spinning blades and the longevity of the machine help determine how much energy you harvest. (Don't know much about kilowatt-hours? Flip to Chapter 4.)

Having good blades is important, but all turbines sold and supported in North America today have serviceable blades that do the job — I'd rarely say, "I won't buy that machine because it has bad blades," though I've seen poor quality on cheap imports.

Depending on the materials, the blades may be the most vulnerable part of your wind generator. Treat them accordingly. Transport them wrapped in protective material, and be careful not to bang them on the ground or against the tower while assembling and installing your wind generator.

The tail and yaw bearing

All small wind turbines turn to align themselves with whatever direction the wind is coming from. A few larger turbines in the *ranch size* (suitable for very large homes or small commercial enterprises) use motors to do this job, but most turn passively, guided by a tail. The wind pushes the tail around, so it's always downwind, leaving the blades upwind. *Downwind* turbines, which are tailless, are an exception, but they still turn into the wind.

The *yaw bearing* is the bearing or bearings that allow the turbine head to turn 360 degrees to follow the wind. These see some wear over a turbine's lifetime, but they generally need attention much less frequently than the main turbine/generator bearings.

The governing mechanism

The idea that you should try to capture all the energy in high winds is a myth. In fact, during storms, a good turbine is working hard to *avoid* capturing all the energy in the wind. All wind turbines worth buying must have some mechanism to deal with times of too much wind. These *governing mechanisms* can take several forms:

- ✔ **Rotor furling:** This involves having the whole rotor turn out of the wind. Some turbines turn the rotor to the side. Others tip it up like a helicopter. And others combine those two motions and furl up and to the side on an angle.

- ✔ **Blade pitch control:** This involves twisting the blades to change their aerodynamics, reducing their exposure and effectiveness in the wind.

- ✔ **Electronic braking:** Braking can slow the rotor and reduce wind loading. This usually happens in conjunction with other governing methods.

- ✔ **Other specialized governing methods:** Other methods include using hinges that flatten or extend blades, blades that deflect wind, parts of blades that change shape, and so on.

The tower adaptor

The *tower adaptor* allows you to connect your specific turbine to the top of your specific tower — which is really great for keeping those spinning blades from becoming your newest piece of lawn art. This component is of steel construction and must connect firmly to both tower and turbine. The tower adaptor is typically a welded steel plate and tube that slips into or over the tower and then bolts on; it then bolts to the turbine via a tube or a flange.

People sometimes overlook this tower-to-turbine connecter in system planning, especially when they buy a tower from one supplier and a turbine from another. Make sure the tower adaptor is included and engineered — you can't make wind electricity without it. (If you buy from separate suppliers, an adaptor usually comes from the turbine manufacturer or from an independent manufacturer, not from the tower supplier.)

The tower

Your tower is one of the most important components in your wind-electric system. Without a proper tower to bring your turbine up where the wind is, your system will perform poorly at best, if it performs at all. In this section, I describe the parts of a basic tower. See Chapter 14 for details on why you need a tall tower and how to choose the best type of tower for your site and situation.

The base and anchors

Except for the simplest of towers for the smallest of turbines, you need a hole for the tower base or bases, where the tower tube or legs land. Your *frost line* (how far down the ground freezes) as well as the tower engineering determine how far down to dig.

For freestanding towers, the base excavation needs to be a big hole to hold enough concrete to keep the tower upright in the face of the winds it'll need to withstand, with no help from guy wires (metal cables — see the later "Guy wires" section). For guyed towers, you need to dig holes for guy anchors as well. Tilt-up towers have four anchor holes, and fixed guyed towers typically have three.

The tower company doing the engineering specifies all hole sizes, so don't start digging without complete excavation engineering. In some cases, you may also need soil analysis, and you always want to know where your frost line is to keep the bottom of the foundation below it. Frost heaves soil up, and if your tower foundations are above the frost line, your tower will be heaved, too, sometimes with disastrous results.

Most holes for wind-generator towers use concrete, though some innovative designs try to avoid this energy-intensive material. You or your contractor needs to be knowledgeable about concrete purchase and installation to make a successful installation. The engineering specs, time necessary to pour, temperature, and other factors affect the concrete mix, additives, and pour procedure.

Within all the concrete is a grid of metal *rebar* — reinforcing rod. The tower engineers also specify the size, number, and configuration of the rebar. You find different steel items in concrete, depending on the tower design:

- ✔ **Inside the anchor concrete:** Here you have large steel pieces running parallel to the ground. Attached to each of these is steel that comes out of the concrete to attach to your guy wires. (The whole assembly — concrete, steel, and steel out of the ground — is the *anchor.*)

- ✔ **Inside the base concrete:** The base concrete contains steel — perhaps threaded — to attach to the tower tube or legs. Sometimes the manufacturers include a base plate that's embedded in the concrete, and other times you attach the tower base or legs after the concrete has cured.

The steel tower parts

Almost all commercial towers use steel as their basic material. This can take many forms (see Chapter 14 for details on these types of towers):

- ✔ **Tubular tilt-up towers:** Sections of steel pipe or tubing, with steel or aluminum couplers that hold the sections together and accept the guy wires

- ✔ **Lattice tilt-up towers:** Sections with tubular or solid steel legs connected with horizontal and diagonal *girts* that tie the legs together

- ✔ **Fixed guyed lattice towers:** Sections of lattice tower, similar to lattice tilt-up towers

- ✔ **Fixed tubular towers:** Sections of steel pipe, welded or sleeved together

- ✔ **Freestanding lattice towers:** Tubular legs and angle iron braces

- ✔ **Freestanding tubular *(monopole)* towers:** Sections of large steel tubing, usually tapering as the tower gets taller

Guy wires

Guy wires (not "guide wires") are metal cables that hold up fixed guyed and tilt-up towers. Without them, your tower ends up on the ground or on your house. Guy wires are either flexible aircraft-type cable or *guy strand,* such as the wire that utilities use to hold up their poles. Guy strand has fewer strands and is quite stiff to work with, though it's effective when installed.

Follow the engineering specifications for guy wires. And if you're making your own tower (see the warnings in Chapter 14), use a substantial safety margin when you size the cables. Don't underestimate the leverage and other forces that the wind exerts on towers.

Transmission wires

Transmission wires deliver the wind electricity to the ground, to your conversion equipment (see the later sections on controllers, batteries, and inverters), and ultimately to the utility grid and/or your home. You need two or three wires to carry the electricity down the tower. They need to be sized to safely carry the maximum energy the turbine is capable of delivering.

Although you should never compromise safety, these transmission wires are frequently sized so you lose some energy in very high winds. Your first priority is to size your wire so it can carry the highest amperage the machine will ever produce, without compromising the safety of the wire and its insulation. But usually this isn't an issue, because sizing to reduce energy loss guarantees the proper safety margin. Sizing wires to lose very little in the extremely infrequent high winds is usually not cost-effective with the price of copper wire today.

Transmission wires frequently run in *conduit,* plastic, or occasionally metal tubing that contains the wires and protects them from physical damage. *Conduit fill* refers to how many wires of a specific size can fit in a given size of conduit; consult your electrician on this.

Transmission wires and conduit typically run down the tower from the wind generator in conduit strapped to the tower or inside the tower tube. They then run underground from the tower base to the electronics room. Local electrical authorities govern the trench depth and other specifications.

Electronics

Wind electricity generators are electrical devices, and downstream from the generator, you have a number of power-conditioning electronic components. See Chapter 15 for more details on all the following components.

The charge controller

If your system includes batteries, you'll definitely have a *charge controller.* This device protects the batteries from overcharge and frequently has a number of other functions, which I discuss in Chapter 15. In many wind systems, a *diversion* charge controller is used in conjunction with a *dump load,* which takes excess electricity and uses it up as heat when there's surplus energy and no grid to sell it to.

Even if you don't have batteries, your turbine will probably have some sort of control box with voltage control devices and metering. I don't recommend buying a wind turbine without metering. You want to know what your machine is doing. It's also important to have a means for stopping the turbine, which is often a brake switch that is included in the controller.

The inverter

An *inverter* converts direct-current (DC) electricity to alternating-current (AC) electricity. DC is what most turbines produce with the wind (technically, your turbine first produces wild AC, but it's changed to DC); AC is what you use in your house and what the grid supplies to you.

In a system that's connected to the grid and that has battery backup, the inverter effectively acts as the charge controller, selling excess energy to the grid to maintain the battery bank at the set voltage when the utility grid is available. In a grid-tied system without battery backup, the inverter sells all the energy that it can to the grid, because it has no batteries to care for.

In a batteryless grid-tied system, the inverter lives and works between the wind generator and grid. In a battery-based system, its home is between the battery bank and the house loads and grid. With higher-voltage wind generators, there may be power conditioning equipment between the wind generator and the inverter or charge controller. This equipment keeps voltage from going too high, so it's often called a *voltage clamp*.

There are two basic types of inverters:

- ✓ **Battery-based inverters:** These inverters need batteries to operate. They work in backup systems, the ones that give you protection from utility outages, and in off-grid systems.

- ✓ **Batteryless inverters:** These inverters convert the wind generator's DC output to AC without batteries involved. These systems provide no backup. When the utility grid is down, you can't make any electricity, no matter how much the wind is blowing.

Choose which type of inverter system you want — battery-based or batteryless — upfront. Changing your mind later is costly. If you're off-grid or want outage protection, your choice has to be a battery-based inverter. If you're grid-tied and don't want or need batteries, buy a batteryless inverter. See Chapter 9 for more discussion of this important decision.

Disconnects and overcurrent protection

All electrical systems need disconnects and overcurrent protection. Every source of energy and every wire must be protected from the type of catastrophic failure that can burn your house down. Disconnects and overcurrent protection are usually built into one common device — the circuit breaker. You'll see a number of these in properly designed wind-electric systems, and they're often grouped in AC and DC boxes. In some cases, fuses are used, though these aren't resettable.

Don't ignore disconnects and overcurrent protection. Although many home-brew systems survive for years with minimal or no protection, following that example isn't wise. You may save a little time and money, but you'll be risking your whole investment and more. And in most places, you'll be subject to inspection, anyway, which requires proper protection.

Batteries

More and more modern home wind-electric systems don't have energy storage. Most people live where the utility grid is fairly reliable, and as long as your system is tied to a working grid, you don't need storage: The utility company provides energy when your system isn't generating enough and takes (and usually credits you for) extra energy you generate. Batteryless systems provide more energy at lower cost, so they're more cost effective and more environmentally friendly.

But some cases call for outage protection — where the grid is less reliable or where *critical loads* (devices that you need 24/7/365) are in place. At other times, having batteries is more a matter of personal preference.

And if you're off-grid, you need battery storage to ensure that your available energy doesn't vary with every gust of wind. (Why batteries? Because energy storage options beyond batteries are pipe dreams, very inefficient, or perhaps the stuff of the future.)

In a system with batteries, the wind generator output goes through the charge controller and into the batteries.

Batteries are often the weakest link in wind-electric systems. They're costly upfront, and they require care and maintenance. Sealed batteries avoid most of the maintenance but are more costly than flooded batteries (which need periodic fluid replacement) and live shorter lives. See Chapter 15 for more details on batteries, battery care, and life with a battery-based system.

Utility interconnection equipment

If you want to connect to the utility grid, you need utility interconnection equipment. Your primary piece of equipment is your inverter, which I discuss earlier in this chapter. It needs to be certified to UL standard #1741 to be legal to connect to the utility grid in most places.

Your inverter is the brains of the interconnection system. It senses whether there's a good quality grid to connect to, and it synchronizes its signal with the grid. If the grid quality drops, it immediately disconnects to keep from feeding wind electricity to a failed grid, which would endanger line workers.

In addition to the inverter, many utilities require specific manual disconnects on the DC and/or AC sides of the inverter. Sometimes these are required to be in a certain location or to be lockable.

Calling for backup: Fuel-fired generators

Batteries have limited storage, so if you're off-grid or want serious backup for utility outages on-grid, you'll likely have a fuel-fired generator to pick up the slack. These are really beyond the scope of this book, and even many renewable energy installers refuse to work with generators. Why? They're a pain in the neck! Generators look easy upfront — you pay your money and get 2 to 6 kW or more of electricity with the turn of a key or the pull of a cord. But after you've lived with and relied on generators for almost three decades (as I have), you realize that this is very expensive, dirty, and noisy electricity. If I could avoid owning, operating, and maintaining a generator, I'd do it in a heartbeat.

If you have an AC fuel-fired generator, you need a battery charger. Note that this is not a *charge controller*, but a *charger* — like what you use to charge your car battery if it gets run down. A high-amperage charger is best so you can get the most out of your generator or the utility grid when you need to charge.

Running a DC generator is one unusual option (I do this) that avoids the need for a charger altogether. These days, most battery-based inverters (see the section "The inverter" earlier in this chapter) have battery chargers built in, so you don't need a charger as a separate component. Running your generator at times of high demand means you can be not only charging your batteries but also running large AC loads at the same time, which is a good use of your fuel and equipment.

Depending on utility requirements and incentive programs, you may also have at least one *production meter* to log kilowatt-hours generated. And you'll undoubtedly have a *revenue meter* to keep track of who owes whom how much.

Getting a Grip on Different System Configurations

How you design and configure your system depends on your goals and situation. This is another crucial decision, determining whether your system gives you the results you want. Here are your system configuration choices and their main advantages:

- Batteryless grid-tied systems are the simplest, cleanest, and most efficient, but they provide no backup.
- Grid-tied systems with backup give you access to the grid along with backup in case of utility outages, but they require battery maintenance.
- Off-grid systems are for people far from the utility lines or those seeking total independence.

In this section, I explain the components that usually go into each system configuration. All of them can be designed to be reliable, long-lasting, and robust. The most important decision is whether you and your values and situations match the system you decide on. Flip to Chapter 9 for full details on these systems, including their advantages and disadvantages and how to select the best one for your energy needs.

Batteryless grid-tied systems

Batteryless grid-tied systems are becoming the most common wind-electric systems installed today. Connecting to the utility grid and avoiding batteries entirely has been a common setup in the solar-electric world for some years, and the wind industry is following suit. Consumers are demanding simple, efficient, and environmentally friendly systems, and these systems deliver.

Batteryless grid-tie systems use these major components (see Figure 3-2):

- ✔ Wind turbine
- ✔ Tower and foundation
- ✔ Transmission wiring and conduit
- ✔ Inverter
- ✔ Power conditioning equipment and dump load
- ✔ Utility interconnection equipment

The wind turbine generates electricity that's pushed down-tower via the transmission wires. Power conditioning equipment may protect the inverter from high voltage, and the inverter takes the DC output and converts it to grid-synchronous AC, to be used in the house and sold to the utility.

Batteryless grid-tied systems are much simpler than systems that incorporate batteries and a backup generator. The sole drawback is the lack of backup capability. When the grid's down, so is your wind-electric system.

Grid-tied systems with battery backup

Grid-tied systems with battery backup give you the benefits of independent systems and the benefits of being connected to the grid. If your circumstances or preferences lead you to want utility outage protection, this is the type of system you want.

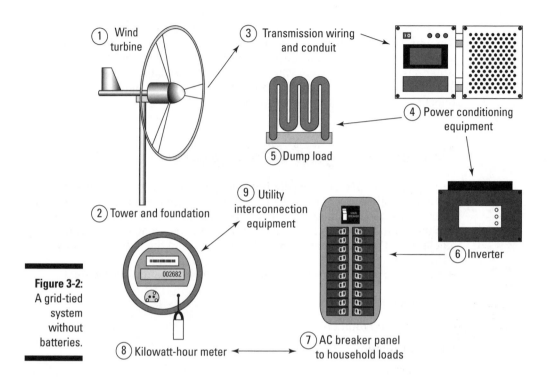

Figure 3-2:
A grid-tied
system
without
batteries.

① Wind turbine
③ Transmission wiring and conduit
④ Power conditioning equipment
⑤ Dump load
⑨ Utility interconnection equipment
② Tower and foundation
⑥ Inverter
⑧ Kilowatt-hour meter
⑦ AC breaker panel to household loads

Battery backup grid-tied systems use these major components (see Figure 3-3):

- ✔ Wind turbine
- ✔ Tower and foundation
- ✔ Transmission wiring and conduit
- ✔ Charge controller and dump load
- ✔ Battery bank
- ✔ Inverter
- ✔ Utility interconnection equipment
- ✔ Possible backup generator

As with the batteryless grid-tied system, the wind turbine generates electricity that's pushed down-tower via the transmission wires. Power conditioning equipment may protect the charge controller from high voltage, on the energy's way to the battery bank. The inverter takes the DC output from the batteries and converts it to grid-synchronous AC for your house or the utility.

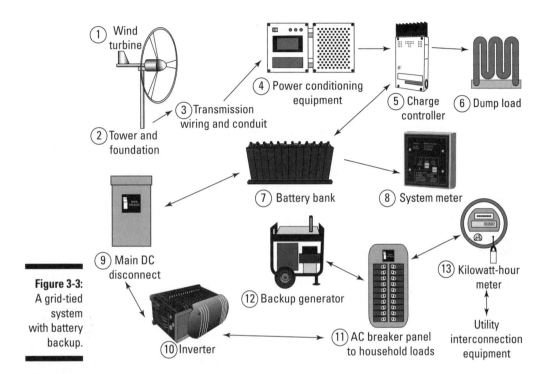

Figure 3-3: A grid-tied system with battery backup.

1. Wind turbine
2. Tower and foundation
3. Transmission wiring and conduit
4. Power conditioning equipment
5. Charge controller
6. Dump load
7. Battery bank
8. System meter
9. Main DC disconnect
10. Inverter
11. AC breaker panel to household loads
12. Backup generator
13. Kilowatt-hour meter

Utility interconnection equipment

With grid-tied battery backup systems, your system doesn't have to provide the entire energy load. The utility grid is there over 99 percent of the time and can pick up whatever load your wind-electric system doesn't produce. Your wind turbine can make only a small percentage of your electricity, if that's what your budget, wind resource, or preferences dictate. But considering your investment in infrastructure, making a large percentage is more cost effective.

Off-grid systems

Off-grid systems use these major components (see Figure 3-4):

- Wind turbine
- Tower and foundation
- Transmission wiring and conduit
- Inverter

✔ Charge controller and dump load

✔ Battery bank

✔ Backup generator

The wind generator on its tower generates electricity that's pushed down-tower via the transmission wires. Power conditioning equipment may protect the charge controller from high voltage, on the energy's way to the battery bank. The inverter takes the DC output from the batteries and converts it to grid-synchronous AC, to be used in the house. A backup generator or other energy sources pick up the slack when there isn't enough wind.

REMEMBER

The wind turbine in an off-grid system is often sized to cover most of the load or at least the windy season load. You need to source the balance of the energy you need from other generating systems, such as solar-electric modules or a backup generator. The battery bank in an off-grid system must be sized for as many hours or days that you want to go without having to start the generator. Inverters and charge controllers are sized to cover the load and highest level of power the systems are capable of generating.

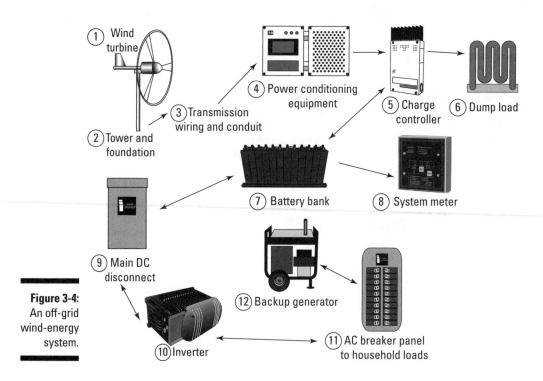

Figure 3-4:
An off-grid wind-energy system.

Chapter 4

Electricity Basics — a Shocking Amount of Info

. . . .

In This Chapter

▶ Understanding wattage and watt-hours

▶ Pushing ahead with voltage

▶ Feeling the flow of amperage

▶ Adding hours to amps

▶ Getting a grip on other nerdy terms and concepts

. . . .

*T*o understand wind electricity, you need to understand electricity in general. As soon as you know the terminology in this chapter and are comfortable using it, electrical concepts should become much clearer to you. You'll be better able to interpret wind-generator and system specifications, and you'll also be able to sort out the hype from the facts in wind-generator advertising.

 Electricity concepts aren't easy! Be patient with yourself if you're new to technical subjects and electricity. Read this chapter a few times over a few weeks, months, or years, and use the terms and concepts in your daily life to get them fixed in your brain. (Hey, starting all your small talk with comments on the weather was getting old anyway, right?)

Getting Up to Speed on Watts

When you want to understand electricity, you need to start with watts. This is the term you'll hear most often, and it's also the term you'll hear most abused. In this section, I explain that contrary to popular belief, a watt is a rate, not a quantity. I also tell you why and how to measure wattage.

Watts (and amps) per hour? Bite your tongue!

Much too often, I hear people say something like, "Does that use 150 watts per hour?" But *watts per hour* is a nonsensical phrase. It's like saying "gallons per hour per hour." Don't say or even think it! A watt is a rate, and you can't talk about a rate over time unless you're talking about acceleration (you're not, so don't go there!). While you're at it, strike *amps per*

hour from your vocabulary for exactly the same reason.

Try to connect *watt* in your mind with *miles per hour* or some other rate. Then when you use the term, ask yourself whether it'd work in place of *miles per hour* in a sentence about travel.

A watt (and a watt-hour, for that matter) is a very small measurement when you're dealing with a modern home's electrical load. A typical American home that doesn't heat with electricity uses between 25,000 and 30,000 watt-hours per day. These numbers get a little unwieldy, so people use shorthand here. *Kilo* is Greek for 1,000, so a *kilowatt* is 1,000 watts, and a *kilowatt-hour* (abbreviated kWh) is 1,000 watt-hours. These terms make talking about energy use much easier.

Understanding that a watt is a rate

If you were to ask me how far Chicago is from Memphis and I answered, "Fifty-five miles per hour," you'd know that something was amiss in my head. You asked a question asking for a distance, and I gave you an answer that's a speed — a rate of distance traveled over time. This sort of confusion isn't common with distances and their appropriate rates because the rates sound like rates. When you say "miles per hour," you know right off the bat that it's a rate of travel that's being discussed — the *per hour* is a big tip-off.

But the term *watt* is shorthand. To me, it sounds like a quantity, such as a pound, mile, acre, dollar, liter, or gallon. It doesn't sound like mile per hour, liter per second, or the like. But in fact, a watt is a rate!

A *watt* (W) stands for "joule per second." I suppose early electrical scientists got tired of saying "joules per second" 18 times a minute and opted for a simple shorthand — watt. A *joule* is a unit of energy, but don't worry about the details here. Focus on the *per* in the phrase. A watt is a rate, just like gallons per second or miles per hour. It's a rate of energy movement. People apply this watt measurement to energy generation, transport, and usage.

The watt is an *instantaneous* measure, just like the reading on the radar gun when the cop catches you going 82 in a 55 mph zone. What he writes on the ticket doesn't tell you how much ground you covered but how fast you were going at the instant he caught you. In other words, wattage describes how quickly energy is being generated, moved, or used *right now*. It can give you info on the following:

✔ The capacity or instantaneous generation of the generator

✔ The rate at which energy is moving through the wires at this instant

✔ The instantaneous energy usage for a particular load, circuit, or home

If your neighbor tells you he paid for 400 kilowatts on his utility bill last month, he's confused. You can't buy a rate! You don't pay for gasoline based on how quickly the pump puts it into your tank, and you don't pay for water based on the speed of the flow. In both cases, you pay for a quantity, not a rate. And with electricity, you buy a quantity — *kilowatt-hours* — from the utility and look at wind generator production in kilowatt-hours. (I discuss watt-hours in detail later in "Watt-Hours: Looking at Energy, the Most Important Measure.")

Another word used for wattage is *power*. Now this may bring clarity to the techies out there, but for most people, it just adds to the confusion. In the non-technical world, people use *power* synonymously with *energy* or with *electricity*. I've given up on fighting the nontechnical use of the word *power*, but at *watts* and *watt-hours* I draw the line — these are two completely different concepts, as different as *miles per hour* and *miles*.

Measuring wattage, the flow of energy

So why should you know the watts — or as I like to say, *wattage* — in your wind-electric system? Well, because wattage describes how quickly you're using energy, it affects everything about your system. If you can use appliances that use energy more slowly but deliver the same results, you use less energy. This means you need to make and transport less energy, which means you need to make and transport fewer dollars to your local wind-energy supplier and utility company.

A 50-watt incandescent light bulb uses energy at the rate of 50 watts (remember, that's a rate — originally joules per second). If you get smart (before or after reading Chapter 7, on increasing your home's energy efficiency) and switch to a compact fluorescent light bulb, you can use about a 13-watt bulb and get the same amount of light. You're using energy at a quarter of the speed, so over the same amount of time, you'll use a quarter of the energy!

You measure wattage using a *wattmeter* (see Chapter 6). How you use this device depends on where you're trying to measure wattage and a bit on what sort of wattage it is. But if you have the standard electricity in most North American homes (120 volts alternating current, or 120 VAC), you can plug a wattmeter into the wall and connect an appliance (like a toaster) into the meter. Set the meter to the *watt* field and watch the reading. You may see several hundred watts or more than 1,000.

Actually doing measurements with a wattmeter can really help you cement the meaning of a watt in your brain. If you plug your clothes washer into a wattmeter, for example, you find that the reading fluctuates a lot. It's very little when the machine is filling, modest when it's washing, and much higher when it's spinning. If you measure your stereo, the wattage is low when the sound is low, and when you crank up the tunes, the wattage increases.

You can also measure wattage on the generating side of your system. You wouldn't generally use a handheld wattmeter in this case, but this function is often built into your charge controller or system-metering package (see Chapter 3 for more on system components). For instance, I can sit at my desk and turn my head 90 degrees to see that at the moment, one of my wind generators is generating energy at the rate of about 1,250 watts and the other is generating it at 1,600 watts — it's a windy day here!

Watt-Hours: Looking at Energy, the Most Important Measure

If I were stuck on a desert island with only one electrical term, I'd choose the *watt-hour.* And if you only have time to figure out one measure, this is it. Why, you ask? The term *watt-hour* (or *kilowatt-hour,* if you have 1,000 of them) speaks clearly about not only how much energy you generate but also how much you transport and, most importantly, how much you use.

All good renewable electricity design is based on first understanding the *load* — how much energy you use. Without good numbers on this end, the whole design is speculation. Before you get excited about that spinning wind generator and see kWh and $ signs dancing like sugarplums, focus on what you really need and how to get it most efficiently (Chapters 6 and 7 lean heavily on the concepts of knowing and decreasing your energy use).

When you know how many kilowatt-hours (kWh) you use — and you've reduced that amount by making your home more efficient — you know how many wind-electric kilowatt-hours you want to make, which can lead you to choosing the right tower, turbine, and system for your home. You may have a battery bank, which you can also quantify in kilowatt-hours, as you can quantify the losses you'll incur by putting energy (kilowatt-hours) into the battery and taking it out. So you see, kilowatt-hours is the most important measure, and it's the one you need to understand more than any other.

Here's where the rubber meets the road. You buy watt-hours from the utility, and when you have a surplus, you sell them back. Watt-hours are a universal measure of electrical energy. If you don't understand watt-hours, you'll have a hard time understanding how to design an effective wind-electric system.

In this section, I provide a formula for calculating watt-hours and give you pointers on how to measure them in your home.

Using a formula for watt-hours

Just in case you think there isn't enough confusion with a watt sounding like a quantity (see the earlier section "Understanding that a watt is a rate"), here comes some more confusion. *Watt-hour* has *hour* in it, so you may think it's like miles per hour or gallons per hour. It's not! A watt-hour is a quantity, like a gallon, a mile, and the like. Sometimes I think the makers of these terms wanted to keep it complicated!

How do you understand a watt-hour? It boils down to simple math, as soon as you understand the concepts. Here's the basic formula:

> Watts (the rate of energy generation, movement, or use) × hours (time) = watt-hours (a quantity of energy)

So using the energy formula, you see that if you leave a 50-watt light bulb on for 1 hour, it uses 50 watt-hours of electrical energy. Leave it on for 2 hours, and it uses 100 watt-hours. In 5 hours, it uses 250 watt-hours, and so on.

I encourage you to try this math out around your home. Use a wattmeter to measure the wattage of some appliances (see the earlier section "Measuring wattage, the flow of energy"), or at least read the labels and use the estimated wattage there. Then time or estimate the number of hours you use each appliance on your list each day. Multiply the wattage by the hours and see what you get. Here are some examples:

- A 1,200-watt toaster running for 6 minutes (0.1 hours) uses 120 watt-hours.
- A 13-watt compact fluorescent light bulb running for 4 hours uses 52 watt-hours.
- A 250-watt heat lamp running for 3 hours uses 750 watt-hours.
- An idle printer drawing 10-watts for 24 hours uses 240 watt-hours.

You get the idea; now do it yourself until the math is easy and the concept is burned into your brain.

Measuring watt-hours

Making up a list of some of your appliances, estimating or measuring time, and doing the math is actually the hard way to figure out how many watt-hours (or *kilowatt-hours* — 1,000 watt-hours) you've used.

Most portable meters that measure watts in your home also measure watt-hours. One version of the Kill A Watt meter, for instance, has a kilowatt-hour button. You plug the meter into a wall socket and plug your appliance into the meter. Then wait until the time (hour, day, week) that you want to measure is over and push the *kWh* button on the meter. It tells you the kilowatt-hours consumed, and another push of the button tells you the number of hours in the measuring period.

Many appliances list watt specifications, but measurement with a meter is always more accurate than using these numbers and calculating. See Chapters 6 and 7 for more on how to make use of watt/watt-hour meters in your home.

Volts: Putting on the Electrical Pressure

All analogies are flawed in some way or another. (I suppose if they weren't, we wouldn't call them analogies but *truth.*) But of all the electrical analogies, I wonder whether calling voltage "electrical pressure" isn't one of the best. *Voltage* is the push that moves electrical energy and electrons from one point to another.

Voltage is also called *potential,* and it's more literally potential *difference* — the difference in electrical pressure between any two points. In any electrical system, you measure voltage at two different points, and the meter compares the electrical pressure at the two points and displays the difference. This is really the same as most any pressure measurement. For instance, when you say that the air pressure in your car tire is 32 pounds per square inch (psi), you're really saying that the difference between the pressure inside the tire and the pressure outside the tire (atmospheric pressure) is 32 psi.

Wind generators create the same sort of potential difference by pushing a magnetic field past coils of copper wire. This voltage makes electrons flow (see the later section "Amperage: Charging Ahead with the Electron Flow Rate" for details). In this section, I define voltage and show you how to measure it.

Understanding what voltage tells you (and what it doesn't)

Voltage describes the push that moves electrons and energy. If you have stronger push, you can move more electrons and energy, and you can do it in smaller wires. Smaller wires means less copper, and copper is expensive. So higher voltage saves you money.

So does voltage tell you everything you need to know about electricity? No way! A voltage reading of 12 volts or 480 volts doesn't tell you anything about the capacity, electron or energy flow rate, or other characteristics. Voltage is just one piece of the puzzle.

Imagine a hose connected to an outside faucet on your home. At the end of the hose is a sprayer trigger. With the faucet on, you water the carrots. You have the capacity of your pressure pump or city water pressure behind the pressure in the hose. When you're done, you let go of the trigger. You still have pressure inside the hose, right? Now turn off the faucet, and as long as your hose doesn't have any leaks, the pressure inside it is still roughly the same. But now the situation is very different: With the faucet off, you have only the small volume of water in the hose itself, and if you hit the trigger, you'll get a small spurt and then nothing — no more flow and no pressure.

Similarly, you can see voltage in a wind-electric system when there isn't necessarily a large amount of energy flowing, or when no energy is flowing. Voltage just indicates the electrical pressure, not the energy flow rate (wattage) or energy amount (watt-hours).

Measuring and interpreting voltage

You measure voltage with a *voltmeter* — big surprise. Put the meter probes in two different points, and your meter measures the difference in electrical pressure between them.

Voltmeters come in a variety of types and sizes. These days, voltmeters are usually digital, which works fine and gives you a high level of accuracy. I also like analog (dial and needle) meters, because watching them through the variations of the wind is so graphic and intuitive. Handheld digital multi-meters (DMMs) always have a voltage function. Figure 4-1 shows a typical multi-meter.

Voltage metering is usually built into wind-electric systems in some place or other, typically in the charge controller or battery state-of-charge meter (see Chapter 3 for more on these components). Here are some common locations for voltmeters:

✔ In a batteryless wind-electric system, you may monitor the voltage of the wind turbine and the voltage of the utility grid.

✔ In a system with batteries, you may see a voltmeter before the rectifiers, showing what the wind generator is doing, and one on the battery, to indicate its voltage.

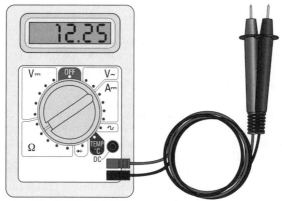

Figure 4-1:
A multi-meter has probe leads to measure voltage.

This section explains how to take voltage readings and interpret them.

Using a voltmeter

If your voltmeter is built into your system, all the smarts you need is enough to read and interpret the numbers. I personally wouldn't buy a system without volt- and ammeters (amp meters) built in, because without them, I can't tell whether the system is working properly.

Voltage features such as polarity (which I discuss in the next section) can help you get wires sorted out properly and troubleshoot systems, so you may have reason to use a handheld voltmeter. Here's how:

1. **Put the meter probe *leads* (the wires) in the proper holes in the meter.**

   Black goes in the COM (black) hole, and red goes in the hole marked with a V (most meters have other notations at this hole — just look for a V).

2. **Before you measure anything, choose between AC and DC voltage (which I explain later in this chapter) and also choose the range of voltage.**

   Voltage choices typically start at 2 volts or below and go to 200 or beyond. Choose a range well above what you expect the voltage to be.

3. **After you've decided where you need to measure, touch one of the probes to one point and one probe to the other.**

   Hold firmly to get a good connection.

4. **After you've made the initial measurement, step down the meter's voltage range to get more resolution.**

Interpreting voltage readings

Touching your probes to the positive (bumped out) and negative (flat) terminals of a flashlight battery is a safe and convenient way to figure out how to use a DMM. If you touch the red probe to the positive terminal and the black probe to the negative, you get a positive number. If you're using a digital meter and you switch the probes, you get the same number but with a minus symbol in front of it.

The difference between negative and positive ends of a battery or another device is called *polarity*. In a wind-electric system, if you have doubts about which side of the DC circuits you're working on (positive or negative), you can check the polarity with a voltmeter. Swapping the leads between the two lines in question won't change the number, but it will add or remove a minus symbol. When that symbol isn't there, your probes indicate the actual polarity — red touches positive and black touches negative. If you're not trying to determine polarity, which probe touches where isn't important — just ignore the minus symbol if you have them swapped.

Voltage interpretation can get tricky when measuring things beyond your flashlight batteries, but remembering that voltage is simply electrical pressure can help. The next step up in complexity may be a simple 12-volt battery bank. If your 12-volt battery bank is fully charged and has been sitting at rest — with no charge or discharge for a few hours — the voltage is about 12.6 volts. If you start using a few lights, the voltage will be pulled down a bit. If you turn on your big 12-volt direct current (VDC) compressor, the voltage may be pulled down considerably, especially when it starts. Conversely, if the wind kicks up and starts charging your battery bank heavily, the voltage may rise to 14 volts or more. All these voltage readings are with a basically full battery bank.

Under a normal usage and charging regimen, a "12-volt" battery in an off-grid cabin cycles from around 12.2 volts to perhaps 14.8 volts. The 12 volts is what's called a *nominal voltage* — it's just the name. In actual service, the voltage varies with the charge and discharge of the battery.

Voltage is not a good indication of how full your battery bank is. Battery voltage increases when the battery is being charged and decreases when it's being discharged. Heavy charge can push the voltage up a lot, and heavy discharge can push the voltage down a lot.

Sometimes you find voltmeters on wind-electric systems that measure voltage directly off the wind turbine, isolated from the battery. This is a useful diagnostic tool, as well as a salve for curiosity. In low winds, you see the turbine generating at lower-than-battery voltages. This means that no energy is going into the battery bank, because you need higher electrical pressure (voltage) to push energy in any direction. As soon as the generator voltage is above battery voltage, charging commences.

Resisting the flow: Ohms

Just as voltage is pressure — the push that makes electrical charges and energy move — *resistance* is the force that slows down the charge and energy flow. The *ohm* (K) is a unit of electrical resistance.

In plumbing systems, if you use very small pipes, you need more energy to push the same amount of water through to your faucet, because there's more friction relative to the rate of water flow. The idea is the same in electrical systems. If you use small electrical "pipes" — smaller wires of copper or aluminum — there will be a strong resistance to the flow of charges. If you use larger wires, resistance will be lower, and charges will flow more easily. Wires must be sized to safely and efficiently carry the charge and energy that the circuit is designed for.

Ohm's Law describes a basic relationship between electrical properties. It states that amperage (A; see the next section) is equal to voltage (V) divided by resistance (R):

$$A = V \div R$$

In plainer language, it says that you can determine how fast the charges are flowing in a circuit by dividing the electrical pressure by the electrical resistance. This formula is used very heavily in electronics but not as much in electrical system design, though being aware of it is a good idea.

Ohm's Law is actually more commonly stated as $E = I \cdot R$, which says voltage (electromotive force, E) is equal to amperage (intensity, I) times resistance (R); I think it's more logical in common situations to state it based on the voltage and resistance, which are commonly more constant and therefore determine the amperage.

Amperage: Charging Ahead with the Electron Flow Rate

Amperage is the flow rate of electrons (unlike *wattage,* which is the flow rate of energy). Amperage is very useful when sizing wire, because wire losses and safety factors are based on amperage, not wattage. And frequently, wind-electric system production is presented only in amps, so you need to apply the power formula (which I discuss later) to get to the more fundamental measure: watts.

If your head can hold the info in the following sections, understanding amperage is good, but don't let this knowledge push out the more important measures of watts and watt-hours. Amperage is really kind of an internal measurement, and by itself, it doesn't tell the whole story. Looking at the amperage is kind of like having a gauge on your dashboard to tell you the revolutions per minute (rpm) of your wheels. With that information — if you know the wheel diameter — you can calculate the speed of the vehicle. You'd need to drive with a calculator stuck to the dash, too, and do quick calcs every time you saw a cop. Having a speedometer on your dash would make a lot more sense (and if some car designer needs to know the rpm, he or she can measure it and think on it). Similarly, if you have a wattmeter, you don't really need an ammeter, unless you're a nerdy designer.

If I were in charge, I'd deemphasize amperage (and amp-hours) and convert as much metering and discussion as possible to watts and watt-hours. A few forward-looking companies (such as OutBack Power Systems with their sophisticated controllers) are moving in this direction, letting their software do the calculation and interpretation so that end users and nerds can directly see the wattage and watt-hours in a system's production and use.

Understanding the flow of charges

My electrical terminology guru, Bill Beaty (www.amasci.com), said something years ago that really helped clear up some fog in my head about electrical terms. He said, "There are two things that flow in electrical circuits — charges and energy."

Understanding what these two flowing things are, and how to talk about them, is critical. Earlier in this chapter, I describe the energy flow rate (watts) and the electrical energy unit (watt-hours). So here, you deal with the other thing that flows, along with its rate and unit. That thing is *charge* or *charges*. (***Note:*** Before I explain, I want to make it very clear that this second set of measurements is less important than watts and watt-hours, and if your head is already stuffed full for today, don't worry too much about this. This measurement is a nerdy, internal, somewhat esoteric measurement, and though it's important to understand it if you'll be the prime designer of your wind-electric system, it's not vital if you're not.)

Charges is a general term for electrically charged particles. In most cases, the term is synonymous with *electrons* — the negatively charged particles in an atom — and that's a clear and simple way to describe them. In other cases, such as in batteries, they're *ions* (atoms that have a negative or positive charge because they have too many electrons or too many protons), but you don't need to get that nerdy here. If *charge* is hard for you, just translate to *electron* when you read it.

These charged particles are actually a part of the circuit — a part of the copper wire, a part of the batteries, and a part of the appliances that electrical energy runs. They're not used up, nor do they enter or leave the circuit. They're part of the materials, just as they're part of your body and all other matter. Drawing from Bill and another mentor, I like to call copper wires "precharged electrical hoses."

When a voltage difference is present in an electrical circuit, the charged particles get pushed along, and their *impulse,* bumping each other along, carries energy. Picture yourself standing in line, waiting to apply for a permit to build your ultra cool wind generator tower. If you push the unsuspecting person in front of you, he'll bump into the person in front of him, and your push will travel down the line. But although you've disrupted a whole bunch of people, you don't get to move very far forward (though you do get to push on the bureaucrat without being fingered . . .).

Note that individual charges, like people in line, move slowly in a circuit, whereas energy moves almost instantly from your light switch to your light bulb when you turn it on. If you had to wait for the electrons at the switch to make it to the light before it turned on, your day would be full of long opportunities for meditation.

Just as energy has a flow rate and a unit, so does charge flow. *Amperage* or *amp* describes the rate of charge (electron) flow. And just as with wattage, this is shorthand. *Amp* is short for coulomb per second. A *coulomb* (C) is a certain number of electrons (6.28 billion billion, to be exact), and an amp describes the number of electrons passing a point in one second.

Just as with wattage, amperage describes a *rate* of charge flow, not the amount! It tells you how quickly charges are flowing at a given instant, not how many have flowed or will flow (I get to the quantity, called *amp-hours,* later in this chapter).

Getting directions: Direct and alternating current

Charges in circuits flow in two distinct modes: direct current (DC) and alternating current (AC). DC and AC don't actually have great differences in efficiency or effectiveness, contrary to some information that's out there. What *is* different is that DC is usually low voltage (low pressure) and AC is usually high voltage. Higher voltage means smaller wire and fewer losses. I explain both modes in this section.

Current isn't a term I'm fond of because again, it doesn't sound like a flow rate, which amperage is. I avoid the term, and I encourage you to at least understand that current isn't a thing that flows in circuits; *current* describes the flow *rate.* Charges (electrons) flow in circuits, as does energy.

Direct current

Direct current (DC) simply means that the charges flow in only one direction. Remember that charges don't join or leave the circuit but are part of the circuit's materials. In direct current circuits, the charges flow through the battery, through the switches, through the lights, back through the battery, and around and around and around. They're not lost or used up but are just bumped along by the voltage (pressure) difference, and their impulse carries energy.

DC circuits include the following:

✔ Low-voltage automotive, marine, and recreational vehicle systems, which are typically 12 volts DC (VDC)

✔ All battery-based systems, because batteries are DC devices; you often see 48 VDC in battery-based renewable electricity systems

✔ Solar-electric (photovoltaic, or PV) modules

Alternating current

The other way charges can flow is called *alternating current* (AC), and this term is very descriptive. The charges in these circuits go not round and round, as in DC circuits, but back and forth. You can think of this as wiggling energy, because the charges just wiggle back and forth. Charges in AC circuits typically reverse direction 100 or 120 times per second, which means they complete 50 or 60 *cycles* per second (see the section "Hertz: Cycles per second" later in this chapter, for details).

You may think that charges in AC circuits don't get much work done, because they just go back and forth, but this isn't the case. Just as a couple pulling an old-fashioned two-person crosscut saw back and forth across a log can transmit their muscular energy into cutting energy, the impulse of charges bouncing back and forth can transmit energy from the source to the load just as easily as charges moving in one direction (DC).

AC circuits include the following:

✔ Standard North American home electricity is 240 volts alternating current (VAC) and 120 VAC.

✔ Many wind generators produce "wild" AC, which varies in voltage and frequency with the rpm of the machine; it's not usable as is. This current is rectified to DC before charging the battery or going to the inverter, which switches the current back to AC (the tame kind).

✔ Utility grids use AC in a variety of voltages, from 120/240 VAC supply for homes to very high voltage (110,000 VAC or higher) lines that run long distances.

Measuring amperage

Measuring amperage with a portable meter is more complicated and hazardous than measuring voltage. When you measure voltage, you're just comparing two points, and you can measure without disconnecting the circuit, with the circuit live. Amperage, on the other hand, most often is measured *within* the circuit — your meter becomes part of the conductor, essentially interrupting the circuit to measure the charge flow (see the later section "Series or parallel: Joining the circuit" for info on how devices that are wired in series become part of a circuit). Though some amp meters (or *ammeters*) clamp on, this, too, requires understanding the circuitry and knowing where to measure and how to interpret the results.

Most commonly in wind-electric systems, you observe amperage via built-in meters, so seeing the amperage from the wind turbine to the loads is easy. These meters are often prepackaged in the wind-electric system controller or inverter.

Ammeters use calibrated *shunts,* metal bars that have a precisely measured resistance. The meter actually measures the voltage drop across the shunt and calculates the amperage using Ohm's Law (I provide a summary of this law earlier in "Resisting the flow: Ohms").

It's common (and safe if done properly) to measure amperage to and from the battery. You're actually measuring the amperage of the charging source and the load in this case.

Never measure amperage across the positive and negative terminals of a battery. All the charges will try to flow from one side to the other because no load is regulating the flow. This will result in at best a blown fuse and at worst a blown meter.

Converting amps and volts into watts with the power formula

Knowing the amperage often isn't very useful in itself, but you can use the amps to help you calculate the wattage. A very basic electrical formula — the power formula — tells you the following:

Volts (pressure) × amps (flow rate of charges) = watts (flow rate of energy)

This gives you some very important information about the relationships between the electrical properties I discuss in this chapter. Here are some examples:

- ✔ 12 volts × 5 amps = 60 watts
- ✔ 48 volts × 1.25 amps = 60 watts
- ✔ 120 volts × 0.5 amps = 60 watts
- ✔ 120 volts × 5 amps = 600 watts

You can't apply the power formula directly to many AC loads because of an esoteric concept called the *power factor* (see "The power factor: A nerdy measure," later in this chapter). But it can be applied to any DC load or to "resistive" loads — incandescent lights, heaters, and the like.

Beyond the math, I hope you'll try to truly understand what the power formula means in conceptual terms. It's tough to explain in plain language — the closest I can come is to say that multiplying the electrical pressure by the rate of charges flowing in the circuit gives you the rate of energy flowing in the circuit. Therefore, if you either increase the pressure or allow the charges to flow faster (by using a larger "pipe"), the energy will travel faster. I hope you can try to grasp this concept, because it can help cement your understanding of the terms and your facility with using them.

Amp-Hours: Knowing the Battery Storage Capacity

With energy movement and storage, you have a rate and a quantity. These are watts and watt-hours, respectively, as I explain earlier in this chapter. With charge flow, you also have a rate and a quantity. The rate is *amperage,* and its unit is an *amp.* The quantity is an *amp-hour* (abbreviated Ah), which is predictable — it's very much like the term *watt-hour.*

An *amp-hour* refers to a quantity of charges (electrons) moved through a circuit. You can't really talk about them being stored, because charges actually just move along whenever the circuit is active.

Watt-hours are a more important and universal measure because they don't depend on voltage, so don't worry too much about understanding amp-hours. But here's one strong reason to understand this measure: Batteries are rated in amp-hours. In this section, I provide a simple formula for calculating amp-hours and explain how batteries are measured in amp-hours.

A formula for amp-hours

As with watts and watt-hours, the math for finding amp-hours is simple:

amps (the charge flow rate) × hours (time) = amp-hours (quantity of charges)

For instance, if your light draws 5 amps and runs for 6 hours, it will have cycled (I wouldn't say *used,* because that may imply *used up*) 30 amp-hours through the circuit. If your pump draws 180 amps for 20 minutes (one-third of an hour), it will have cycled 60 amp-hours.

Looking at amp-hours as a battery measure

Batteries are rated in amp-hours, so you need to have at least a basic understanding of the measure if you plan to have a wind-electric system with batteries. In fact, as I discuss in detail in Chapter 15, there's more to this rating than meets the eye. A typical deep-cycle battery for a wind-electric system is rated in amp-hours "at the 20-hour rate." What does this mean?

A battery's charge and discharge capacity isn't a fixed number based only on its size, the amount of lead in it, or other battery characteristics. It's also based on how quickly you're charging or discharging it. When you remove energy from a battery very quickly, it's not very efficient, so the losses can be quite high. If you remove energy more slowly, you get more total energy out of the battery.

A *20-hour rating* is the amount of charge that you'd cycle if you were to remove all the battery's energy over a 20-hour period. So for a battery with a 100 Ah capacity, the 20-hour rating would mean charging or discharging at a rate of 5 amps. In the real world of home renewable-energy systems, you often cycle at a much slower rate, perhaps even a 100-hour rate. So you actually see more energy storage and release capability in these systems than the 20-hour rate.

Rating batteries in amp-hours is unfortunate in my opinion, though there are some nerdy reasons that perhaps pushed this rating system. I'd prefer to see batteries rated in watt-hours and encourage you to do some simple math to get there from the rating:

Battery voltage × amp-hours = watt-hours

Measuring amp-hours

Amp-hour meters aren't uncommon, and this measuring device is the basis of battery state-of-charge meters. These meters are "bean counters," noting the passage of charges in both directions, applying an efficiency factor, and calculating the charge level of your battery. These require careful setup and calibration.

Sophisticated handheld meters can also log amp-hours. But again, using a meter that measures watts and watt-hours instead can simplify your life and give you the information you're really after. It can also spare you from using the calculator so much.

Putting It All Together with a Handy Electrical Analogy

Over years of my own study and understanding of electrical terminology, I've seen people make many analogies to try to increase understanding of the terms. But all the analogies I've seen break down in serious ways at some point or other. For instance, many analogies use water flow rate to represent both watts and amps. This may work if you're talking about the terms independently, but it doesn't work if you're trying to understand the difference between the two.

For years I longed for a unified analogy that described the basic electrical terms and their relationships to each other. In the end, I had to make my own. I'm not suggesting that mine can't be picked with (I welcome your suggestions for improvement), but I think it does a good job of showing what each term means, comparing electrical ideas to concepts that most people understand.

Any unified analogy needs two flowing things to represent the two flowing things in electrical circuits: charges and energy. I chose to use a water-powered grain mill in which water represents the charges and the grain and flour represent the energy. Check out the terms, and see Figure 4-2 for some visuals:

- **Water pressure represents voltage, the electrical pressure in the wires.** The system actually has two sources of pressure: the pressure from the pump and the gravity pressure from the tank on the hill. These are the motive forces that make the water move in the system.

- **Electrical resistance, measured in ohms, is similar to resistance in a water system.** All pipes, fittings, and water-powered loads slow the flow of water. In electrical circuits, the wires, connections, and loads do the same for charges.

- **The water flow rate represents amperage.** Note that this is a closed system — the pipes (wires) are precharged with water, and water is never added or lost. It just goes around and around and around in this direct current (DC) system. With the proper mill, you could do the same work with water that switched directions frequently, like AC electricity.

- **Amp-hours are like the quantity of water that passes a point in a given time.** You could sit with a flow meter and a stopwatch and calculate how many gallons have passed a point in a minute. Remember that amp-hours don't represent *energy* but a quantity of charges that have moved.

- **Wattage is the speed at which the grain is ground into flour.** Remember that wattage isn't a quantity of "stuff" but the rate at which stuff (energy) is generated, moved, or used.

- **Watt-hours are shown in the pile of ground flour.** They're the actual work done.

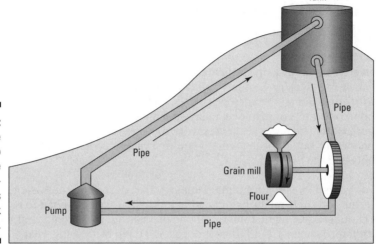

Figure 4-2:
A simple analogy to describe how electrical terms work together.

My hope is that this analogy helps you understand the key electrical terms and how they relate to each other. Here are the main lessons:

✔ Two things flow in electrical circuits: charges and energy.

✔ Each of these has a flow rate (speed) and a quantity (amount).

✔ Amps and watts are flow rates, not quantities of stuff!

✔ Amp-hours and watt-hours are quantities of stuff — in this case, charges and energy.

Checking Out Additional Electricity Terms and Concepts

I discuss what I think are the most important electrical terms earlier in this chapter, but of course, there are many more. In this section, I briefly outline some of these other less-important terms so you at least have some familiarity with them.

Series or parallel: Joining the circuit

For charges to flow, they need a closed path, or *circuit*. You can create a single loop that goes through all your devices, or you can have wires branch off and offer the electricity multiple ways to complete its path.

Series refers to wiring electrical devices (photovoltaic [solar-electric] modules, batteries, loads, and so on) in a string, with each unit like a link in a chain. When you stack four flashlight batteries in your big flashlight, for example, you're wiring them in series, connecting positive to negative and positive to negative all the way along.

Series connections increase the voltage without altering the amperage (or amp-hour capacity, in the case of batteries). Therefore, your four 1.25-volt batteries wired in series give you a 5-volt flashlight. Here's what you may see wired in series:

- ✔ Wiring batteries and photovoltaic (PV) modules in series to attain higher voltages is quite common.

- ✔ Switches are wired in series with the specific load you want to switch on or off. All safe circuits have some sort of overcurrent protection and disconnect ability, usually both built into a circuit breaker. These are wired in series with the circuit so that all charges flow through them; that way, you can completely disconnect the circuit manually or have it disconnect automatically in the case of a fault.

- ✔ Ammeters and wattmeters are also wired in series, because you're trying to measure the flow of charges and energy in a specific part of the circuit. (See the earlier sections "Measuring wattage, the flow of energy" and "Measuring amperage" for more on these measuring devices.)

In a series string, each device is dependent on *all* devices being operable to pass charge and energy. As with old-style Christmas lights, if one bulb goes out, the whole string dies. For this reason, you rarely wire loads (such as your waffle maker and refrigerator) in series.

Parallel is the opposite of series, and it describes wiring devices so that each is separately connected to the energy source. Items wired in parallel retain the same voltage but increase amperage (or amp-hour capacity in batteries). If one wire breaks, the charges and energy can take a different route, so your devices keep working. Here's what you may see wired in parallel:

- ✔ People wire almost all loads in parallel.

- ✔ *Strings* of batteries or PV modules are also often paralleled.

- ✔ Voltmeters are wired in parallel with a circuit; they touch and sense the electrical pressure only at the specific points of contact, without charges flowing through the meter. (See "Measuring and interpreting voltage," earlier in this chapter, for more on voltmeters.)

Hertz: Cycles per second

In alternating current (AC) circuits, the flow of charges changes direction many times a second (see the earlier section "How charges flow: Direct current and alternating current"). People typically describe this change by identifying the number of cycles per second.

A *cycle* is a complete trip from a complete halt to movement in one direction to a complete halt to movement in the other direction and back to a complete halt. So in fact, in one cycle, the charge-flow direction changes twice. The term *hertz* (abbreviated Hz) describes 1 cycle per second.

In North America and many other parts of the world, electricity is 60 hertz. In much of Asia and Europe, it's 50 hertz. The primary practical use for understanding this term is to use appropriate appliances and electronics with appropriate electricity sources. For instance, if you're living in North America, you buy a 60-hertz blender; if you're in Europe, Asia, or Africa, your blender will likely be 50 hertz.

At the tower top, most wind turbines produce AC with variable frequency (and voltage). When the rotor spins faster, the frequency increases. This variable-frequency and variable-voltage AC isn't usable, so it's converted to DC, often right at the alternator.

The frequency of your local utility grid determines not only the frequency of your appliances but also the frequency of your inverter, which changes the DC electricity back to AC. So if you live in North America, you'll be buying appliances that run on 120/240 VAC, 60 hertz, and your inverter will have to meet the same specifications.

The power factor: A nerdy measure

Ideally in AC electricity, the voltage and amperage (the pressure and flow of charges) would peak at the same time. But properties called capacitance and inductance (together called *reactance*) slow the flow of charges and can send the voltage and amperage out of sync. When this happens, your generator needs more capacity because some of the charges are sloshing about but not delivering. There's not a great loss here because the extra charges and energy return to the source, but you temporarily need more generating capacity.

Power factor defines the level of sloshing and the extra generating capacity you need. A 0.8 power factor means your generator needs to be 20 percent larger to support the load, so a 10-watt light bulb would need a 12-watt generator. The bulb still *uses* only 10 watts, but it needs to have 12 coming in and essentially sends 2 back.

Many appliances, including most electronics, fluorescent lighting, and the like, don't have a perfect (1.0) power factor. For homeowners, this isn't a big deal, because the utility charges for watt-hours — the actual energy used — regardless of the power factor. For a wind-electric system, power factor does not usually cause any large design changes, though you may size your inverter somewhat larger if your battery-based home has many loads with a low power factor.

Chapter 5

Blowing through Vital Wind-Energy Principles

. . . .

. . . .

*U*nderstanding the physics of wind energy is essential to making good system design decisions. The concepts in this chapter aren't rocket science, but wind energy also isn't what too many people think it is — easy, free energy. Understanding more about wind resources (see Chapter 8) and about system design (see Chapter 16) can help you take this chapter's basic principles about wind energy and put them to use. This realistic and practical grounding can lead to a successful system that actually makes usable electricity — and exceeds your expectations.

Understanding Wind Speed Terminology

People's casual observations of the wind are very unscientific: "It's been windy all week." "It was dead calm all day." "It never stopped." And when casual observers start throwing around numbers, they're even more unscientific: "It's always blowing more than 20 miles per hour here." "We saw gusts of 80 miles per hour." "It's been a steady 15 miles per hour."

All these statements are suspect when you try to evaluate the *wind resource* (how much wind energy potential you actually have) on a site. They make for fun conversation, but that's all. Toss them in the trash can if you're serious about evaluating your wind resource in a meaningful way. One important step in such serious evaluation is getting a handle on wind speed terminology. In this section, I define three types of wind speed: instantaneous, average, and peak.

In the utility-scale wind world and in many countries around the world, *miles per hour* (mph) isn't the standard. Instead, people use *meters per second* (m/s) as a normal metric measure of wind speed. In home-scale wind systems, using miles per hour (mph) is common, though you may hear the pros using meters per second. To convert, know that 1 m/s = 2.24 mph and 1 mph = 0.45 m/s. If you know meters per second and want miles per hour, multiply by 2.24. If you know miles per hour and want meters per second, divide by 2.24 (or multiply by 0.45).

Instantaneous wind speed

Wind is a *variable* resource — it rarely stays the same for moments, let alone minutes. In Chapter 4, you find out that a watt is an instantaneous measure of energy generation, flow, or use. It gives you information about right now, not before, later, or any cumulative info. *Instantaneous wind speed* is the same — it's a right-now measurement of the actual wind speed, with no history, no average. This means that the instantaneous wind speed varies all over the map.

Knowing what to expect in terms of instantaneous wind speeds on a property is interesting but not particularly useful. Instantaneous wind speeds obviously start at 0, which is dead calm. And depending on your specific site, they may go up to 100 mph or more. Non-storm winds on most sites vary from 5 to 35 mph. By the time you get up to gusts of 35 mph, you know it's blowing! Wind system designers have little or no use for this casually observed information. You could sit and watch an instantaneous anemometer all day long and still not have a real idea of your wind resource.

Your crucial measure: Average wind speed

An instantaneous wind speed doesn't give you the information you need to evaluate a site or predict output from a wind generator. Instead, you're looking for an *average* wind speed. The average wind speed tells you how much overall wind your turbine will see and therefore how much energy it'll generate.

To get an accurate average wind speed, a device looks at all the instantaneous wind speeds over a specific period of time and computes an average. For instance, suppose you have a site that blows 12 mph half the time and 8 mph the other half of the time (again, this isn't realistic; wind is variable). The average wind speed on this mythical site would be 10 mph.

Wind site analysis for utility-scale wind farms is much, much more complicated. Prospectors look at wind distribution (the percentage of time at each wind speed) and plot available energy in detail. For home-scale systems, a simple average wind speed is adequate for predicting wind generator production. I explain more about finding average wind speed in Chapter 8.

Average wind speeds on sites that humans inhabit range from the low single digits to about 15 mph average. The top end of that range is actually quite rare for home-scale wind sites. Installing home-scale wind generators on sites with 7 to 12 mph average wind speeds is common. Below 6 or 7 mph average wind speed, you don't get a lot of energy (though off-grid folks may still want to tap winter winds), and above 12 mph, you can't find many sites.

Peak wind speed

In addition to instantaneous wind speed and average wind speed, you want to know the peak wind speed on a site and, ideally, how frequently it occurs. The peak wind speed gives you some indication of the worst-case scenario for your wind turbine — the highest forces it'll need to withstand.

Many wind turbine manufacturers say that their equipment is rated to operate in up to 120 mph winds. I'm not sure how they know this — I doubt they've tested the turbines in real-world conditions. I suspect that it's an educated guess, and I don't put very much stock in this number, especially because most manufacturers seem to use the standard 120 mph.

What I do put stock in is that high winds pack a punch because of the cube of the wind speed (see the upcoming section "Wind speed cubed (V^3): A dramatic effect"). How heavy-duty your wind-system needs to be depends in part on how severe the winds on your property can be. For example, your wind turbine has to deal with more than 200 times as much energy in a 60 mph wind as it sees in a 10 mph wind. So if you tell me that your site never sees 60 mph, I might recommend a different turbine than if you tell me that you see 80 mph five times a winter.

Gaining Lessons from a Basic Wind Energy Formula: $P = \frac{1}{2}DAV^3$

A basic wind energy formula provides crucial lessons about tower height, turbine size, and user expectations. This formula isn't theory but basic physics. It can help you understand the fundamental principles behind some important design decisions, and it holds more sway than all the marketing hype from wind turbine designers, salespeople, and scam artists.

The formula is a condensed version of a very complex formula that I'll spare you (and myself). The condensed formula summarizes the power available to a wind turbine in this way:

Power = $\frac{1}{2}$ × the density of the air × the swept area of the wind generator × the wind speed cubed

In short, $P = \frac{1}{2}DAV^3$. Note that the formula is not talking about what a specific wind generator will do with the available wind power, just the general potential. Designs and efficiencies vary, and this formula isn't designed to specifically predict power or energy output. I use it here as a demonstration of principles.

Later in this chapter, I do give you some tools to estimate production from specific turbines, including a couple of formulas. In the following subsections, I break down what the formula means for a wind system.

Power (P): What's available

Technically, power and energy aren't the same thing. In common speech, people use these terms interchangeably. But if you're with a group of electricity or physics geeks, *power* means watts and *energy* means watt-hours. *Wattage* is the rate of energy generation, movement, or usage. *Watt-hours* are actual energy units — the capability to do work. (I explain the difference between the two concepts in detail in Chapter 4.)

In other words, power is an instantaneous measure, not a cumulative one. Power describes the energy available, or *potential,* in one instant, not the total amount of energy after a certain length of time.

The formula $P = \frac{1}{2}DAV^3$ is talking about the power available in the wind, at a particular air density and speed, as it hits a rotor of a certain size. The *rotor* is the wind generator's blades and hub, so its size is pretty much constant. But wind speed changes all the time, so power can vary from moment to moment.

Air density (D): A hard-to-sway factor

The first factor on the right hand side of the equation $P = \frac{1}{2}DAV^3$ is the air density. Air density is a physical fact to be aware of, but it has no action agenda for you. It's not something that usually changes purchasing or design decisions, unless you're installing at high altitudes.

Air density may have a significant impact on the power available to your wind turbine. When temperatures are cool, the air is denser. Because moving air is your fuel, moving dense air gives you more available power than moving thin air. When it's hot, the air is thinner, so less power is available to your turbine. So winter winds have more power in them, and summer winds, less. The variation between summer and winter is hard to measure, and there's no practical way to use the information, anyway — you'll simply enjoy a bit more production in the same winds when it's cooler.

Higher altitude means thinner air as well, whereas air at sea level is much denser. A standard factor is to derate your wind turbine's estimated production by about 3 percent for every 1,000 feet above sea level. Table 5-1 shows a chart of this altitude correction.

Table 5-1	Turbine Performance Changes with Altitude	
Elevation (Feet)	*Elevation (Meters)*	*Relative Performance (%)*
0 (sea level)	0 (sea level)	100%
1,000	305	97%
2,000	610	94%
3,000	915	91%
4,000	1,220	88%
5,000	1,524	85%
6,000	1,829	82%
7,000	2,134	79%
8,000	2,439	76%
9,000	2,744	73%
10,000	3,049	70%

Elevation can be a big deal. For example, if I install my turbine at sea level (well, 170 feet above sea level) at my island home and you install yours at 10,000 feet in the Colorado Rockies, you'll get 30 percent less energy than I do. Beyond being aware of the differences, you can't do anything about this factor in the equation, beyond buying a turbine with a larger diameter if you're going to install it at a high altitude.

Swept area (A): Collector size matters

Wind journalist and colleague Paul Gipe wrote, "There's nothing that affects performance of a wind turbine more than its swept area, except the wind resource itself." *Swept area* is the circle that the collector surface of a wind turbine — the spinning blades *(rotor)* — describes. The collector is what actually captures wind energy. Behind the turbine's rotor is the generator, which converts the wind energy into electricity.

Wind energy newbies seem to focus on the wind speed, on the generator size, and on other parts of the system or design. In the end, though, the auto gearheads have it right: There's no replacement for displacement (in their case, it's cylinder displacement; you can't make a super powerful car with tiny cylinders).

If you buy a solar-electric module or solar hot water collector, you expect a certain amount of energy from it. If you buy a second unit of the same size, expecting double the production is quite logical. The idea is the same with a wind turbine's swept area: The bigger your collector, the more energy you collect.

If you double the collector size (and therefore double the swept area), you double the power potential. Don't overlook this fact. No magic behind the blades can make lots and lots of energy when there isn't much swept area driving it.

Interpreting swept area

Swept area is commonly identified in square feet or square meters. Other times, however, people talk about turbines by their diameter in feet or meters. This is handy shorthand, but understanding how diameter relates to available power isn't terribly intuitive.

Check out the basic math: You determine the area of a circle by multiplying the square of the circle's radius (half the diameter) by π, where π is about 3.14. In other words, $A = \pi r^2$. Look at how collector area increases as you increase the diameter:

- **3-foot diameter:** $3.14 \times 1.5 \times 1.5 = 7$ square feet
- **6-foot diameter:** $3.14 \times 3 \times 3 = 28$ square feet
- **12-foot diameter:** $3.14 \times 6 \times 6 = 113$ square feet

As you can see, when you double the diameter, the swept area actually quadruples (Figure 5-1 gives you the corresponding visual to help you calibrate your guesser). To double the swept area, you have to increase the turbine's diameter by only about 41 percent; to triple it, just increase the diameter about 73 percent.

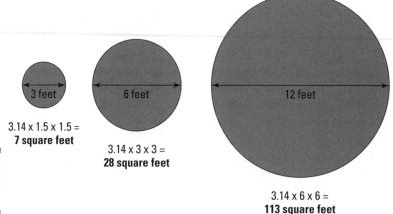

Figure 5-1: Calculating swept area.

3 feet

$3.14 \times 1.5 \times 1.5 =$ **7 square feet**

6 feet

$3.14 \times 3 \times 3 =$ **28 square feet**

12 feet

$3.14 \times 6 \times 6 =$ **113 square feet**

Some wind turbine manufacturers have wisely included the swept area in square feet in the model numbers of their turbines. For example, the

Abundant Renewable Energy (ARE) 442 is roughly 442 square feet, whereas the company's smaller turbine, the ARE 110, is about 110 square feet. Comparing turbines by square footage of swept area makes a lot of sense. Again, this footage is the collector area — don't underestimate it.

Passing on tiny turbines

I frequently read about inventors coming up with "new" discoveries of turbines to mount on your roof (don't do it!) or in your yard. These very often are tiny turbines with a diameter of 1 to 4 feet, but the inventors or marketing companies make large claims. Don't believe them. You don't expect a 40-foot yacht to go very fast if you put dinghy sails on it. In the same way, don't expect a tiny micro-turbine to produce very much energy — it has a tiny swept area, so it intercepts only a small amount of wind energy.

Typical wind turbines for modern homes are generally in the 12- to 50-foot diameter range. Smaller than this may provide some energy for a super-efficient home in a super-windy environment or in combination with other energy sources. But be realistic about the potential available to your turbine.

Wind speed cubed (V^3): A dramatic effect

The wind speed itself is the biggest factor in the power formula, $P = \frac{1}{2}DAV^3$. And though it may seem like you have no control over wind speed, that's not the case. Your choice of site and tower height can affect the wind resource dramatically. See Chapter 8 for details on how to determine your site's potential and how to maximize it. In this section, I define *cube factor* and show you how it affects power at a range of wind speeds.

Understanding a cube factor

Double the force on your bicycle pedals, and you can expect to go twice as fast; put twice as much coal in your furnace, and you'll make about twice as much heat. Those relationships are linear. But the power available in the wind relates to the *cube* of the wind speed (V^3, or $V \times V \times V$). The upshot is that small changes in wind speed can mean big changes in power. Take a look at some various wind speeds cubed (I'm thinking miles per hour here, but the units aren't critical):

- **3 mph:** $3 \times 3 \times 3 = 27$
- **6 mph:** $6 \times 6 \times 6 = 216$
- **8 mph:** $8 \times 8 \times 8 = 512$
- **10 mph:** $10 \times 10 \times 10 = 1,000$
- **12 mph:** $12 \times 12 \times 12 = 1,728$
- **20 mph:** $20 \times 20 \times 20 = 8,000$

A doubling of the wind speeds gives eight times the available power. And at some points, gaining 2 miles per hour doubles the available power. (Of course, that works the other way, too: Cutting the wind speed in half gives you only an eighth of the power. That idea can help you look at power curves with a critical eye, as you see later in "Dangerous curves: Why power ratings are misleading.")

The farther you get away from the Earth and its natural and artificial obstructions, the higher wind speeds you have. Gaining a few miles per hour by adding 20 to 60 feet to your proposed tower height is relatively easy. And why not? Short towers shortchange the performance of the systems. In fact, installing a short tower is the the most common mistake in wind-electric system design.

What a cube factor means in low winds

The cube factor is unimpressive at low wind speeds: $2 \times 2 \times 2$ is only 8, compared to $10 \times 10 \times 10 = 1,000$. Even $5 \times 5 \times 5$ is only 125. The lesson here is that very little energy is available in low-speed winds.

Here's an example: I lived for years with a couple of wind turbines that had very different characteristics. One was large and heavy and took a significant wind (perhaps 9 mph) to start up, though it would generate a bit at 7 mph after it started. The other would start up in very low winds (perhaps 4 mph) but actually didn't generate much energy until 7 or 8 mph, as is normal for most turbines. I'd routinely have visitors who'd be excited to see the latter turbine spinning, unaware that it actually wasn't doing anything beyond wearing out its bearings and attracting attention. I actually preferred the heavier turbine, because it didn't start spinning until there was really something to capture.

Anyone who brags about wind turbine performance below 7 or 8 miles per hour is blowing smoke. There just isn't much energy to capture down there. When putting together a wind system, don't give a second thought to what your wind turbine will do for you in winds that are less than 10 mph instantaneous — the energy there is insignificant. And get your machine up on a tall tower so you can get into the real winds.

What a cube factor means in high winds

What about cubing high wind speeds? For instance, how about $40 \times 40 \times 40 = 64,000$? Or $50 \times 50 \times 50 = 125,000$? These numbers are huge, and they indicate the relative power available in winds of these various speeds.

A common first reaction when thinking about high winds is, "I want to capture all of it." But the wind turbine engineer doesn't come to that conclusion. Instead, with these kinds of forces, you have to *protect* the machine from the wind so the machine can live another day. Trying to capture the energy in high winds would mean designing an extremely strong machine. This would likely mean a very heavy machine, and this machine likely wouldn't perform well in the moderate winds that you see most of the time. See Chapter 13 for detail on this subject.

Don't let concern about high wind speed cause you to choose a shorter tower. Shortening a tower reduces overall production, and the machine will still see high winds on occasion. Every machine worth buying needs to be designed to *govern,* or avoid capturing wind energy, in high winds.

See Chapter 3 for info on governing mechanisms, and see the earlier section "Peak wind speed" for info on the maximum speed on a site.

Where's the sweet spot?

The wind-speed distribution curve in Figure 5-2 is important to understand. Look at the frequency of winds on the low end — you don't see very low speed winds all that often. Look at the frequency of winds on the high end — these winds occur even less frequently. So if you're not focusing on very low or very high wind speeds, that leaves the middle to attract your gaze. And in fact, that's the important range of wind speeds to be concerned about. The range of roughly 10 to 20 mph is where most winds occur on most home sites. So this is where you want to have good production from a wind turbine.

How often do people actually see high winds? On a typical wind site, it's a very small percentage of the time, like less than 2 percent. Designing a machine to get all the energy out of those infrequent storm winds just doesn't make sense.

If you hear companies or inventors bragging about performance of a wind turbine at 3 to 7 mph or talking about *startup* (the speed at which a turbine starts spinning), they likely don't really understand the physics of wind or are trying to pull wool over your eyes. And if they brag about performance beyond about 25 mph, they're trying to fool you, or they haven't actually designed a system that protects the turbine in high winds. Though the turbine may initially give impressive results in high winds, the end result in most cases will be damage — and dead turbines produce no energy.

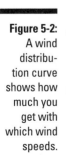

Figure 5-2:
A wind distribution curve shows how much you get with which wind speeds.

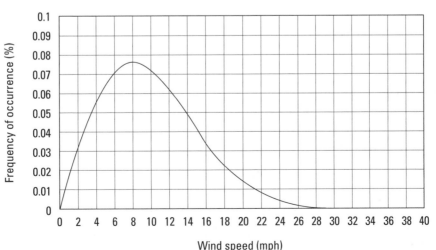

Knowing that Energy, Not Power, Is Important When Predicting Output

Whether you use the terms *power* and *energy* technically or not, understanding the conceptual difference between the two is vital. Power (watts) is an instantaneous view of turbine production — the rate of energy production — and energy (watt-hours) is the cumulative generation.

Unfortunately, power seems to get a lot of press, whereas the much more important energy is often overlooked. Wise consumers look at things the other way around — you find out why in this section.

Dangerous curves: Why power ratings are misleading

Often, wind generator manufacturers use power curves to market their machines. A *power curve* shows predicted or measured wind generator output across the full range of wind speeds (see the earlier section "Instantaneous wind speed" for more on varying wind speed). See Figure 5-3 for an example. It shows that at 8 mph, the example machine generates about 50 watts; at 15 mph, about 350 watts; and at 30 mph, about 1,200 watts.

When looking at the curve, most people's eyes tend to land at the top end: the *peak watts*. It indicates the highest instantaneous production that you'll see out of a turbine. In Figure 5-3, it's 1,200 watts. So what does this tell you? Very little.

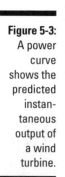

Figure 5-3: A power curve shows the predicted instantaneous output of a wind turbine.

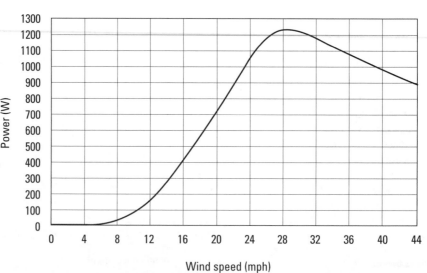

Manufacturers rate wind turbines at a certain wind speed. This *rated power* — typically the same as peak watts — is a very misleading number. It doesn't predict the energy generation, and it tells you little about average or total production. Here's why:

- ✔ **Inflated wind speed estimates:** The rated wind speed is often in the 25 to 30 mph range, but sometimes it's higher: In rare instances, it's as high as 70 mph. Even the low end of this range is well above the average on all home sites — roughly 10 to 20 mph — and well above where most winds occur. If a turbine is rated at 28 mph, for instance, its actual power at 14 mph (a far more typical wind speed) is one-eighth of the rated power (due to the cube factor that I discuss earlier in this chapter); that means a 1,000-watt turbine would really be a 125-watt turbine in the winds that are most common on most sites!

- ✔ **Variations in wind speed:** Winds on all sites are variable, but you actually see the rated wind speed only rarely. Think about it this way: With your backup gas generator, you keep getting the rated output as long as you keep it gassed up. Your wind generator would do the same if it were supplied with a steady diet of the rated wind speed (for example, 28 miles per hour). Will your wind turbine see steady winds? Not at all.

So what can you glean from a power curve or power rating? Almost nothing! (Hey, haven't you always heard that power corrupts?) Well, that may be an exaggeration, but most people don't have the background to interpret a power curve and end up with usable information for a turbine buying decision.

Estimating output: Why energy curves are better indicators

When you buy electricity from the utility, you pay for kilowatt-hours (kWh). This is the measure you should focus on when looking at a wind turbine, too. Ignore the peak or rated power (which I discuss in the preceding section), and find out how much energy you'll get with the turbine at your site.

To get the energy number, you need to know your average wind speed at the height you intend to install your machine. (See the earlier section "Your crucial measure: Average wind speed," and flip to Chapter 8 to find out how to get this number.) Then look at manufacturers' production estimates to get an idea of what each machine will do for you. Each company (at least the ones worth doing business with) has energy curves, graphs, or tables showing the estimated energy production at a variety of average wind speeds, from about 8 mph to about 14 mph. Figure 5-4 gives an example.

An energy curve or graph shows not instantaneous power but energy over a certain time period. Some manufacturers show monthly data, and others show annual. Looking at the example table in Figure 5-4, you can see the monthly production for a 12-mph wind speed and this company's 48-volt machine is estimated to be about 220 kWh.

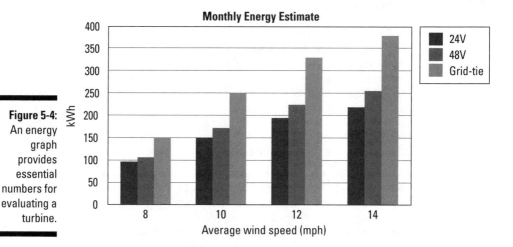

Figure 5-4: An energy graph provides essential numbers for evaluating a turbine.

Suppose your site assessment tells you that your average wind speed at 120 feet on your site is 9 mph. Your budget is holding you to a smallish machine, and Bergey Windpower's XL.1 is within your price range. According to the Bergey Web site, the XL.1 will produce about 85 kWh per month on an 8.9 mph average wind speed site. That's about 3 kWh per day. This gives you a solid estimate to work from in your system design, budgeting, and planning (I provide the details on these steps in Parts II and III).

Compare that to trying to make sense of the same turbine's *rated power,* or its maximum power. Bergey rates this machine at 1,000 watts at 24.6 mph (11 meters per second), and it peaks at about 1,200 watts (1.2 kilowatts). But from those numbers, it's nearly impossible to calculate or extrapolate what you'll actually get on your site with your winds.

Here's another example of how misleading rated power can be. I lived for several years with the two following machines:

Machine	Rated Power	Turbine Diameter	Swept Area
No. 1	1,000 watts	9 feet	64 feet²
No. 2	900 watts	12 feet	113 feet²

If you were to look only at the rated power of these two turbines, you may reasonably think that Machine No. 1 was more productive. But look again. What about the swept area? Machine No. 2 has close to twice as much collector area, so I'd expect it to produce more energy (as I explain earlier in this chapter). In fact, my ongoing kilowatt-hour (kWh) measurements showed that Machine No. 2 produced 2.3 times as much energy as Machine No. 1.

If you're price shopping, look at the swept area — square footage or square meters — of the machine (in addition to the energy numbers); then see what you'll have to pay per area of collector. Do not compare turbines based on the price per rated watt of output. This will lead you to buying a lightweight, light duty turbine that won't last long if you have a medium or heavy-duty site.

As one prominent wind turbine salesman said at a national conference, "Peak power is a marketing number." You're looking to generate energy, so keep your eye on that number: kilowatt-hours.

Balancing energy predictions with truth formulas

Because the wind power industry doesn't yet have standardized testing (or even a standard to test to, but that's coming), you generally have to rely on manufacturers' estimates of energy production from each machine. If you don't have these numbers (or don't trust them), having some supplementary methods to estimate production is useful.

Two prominent wind experts have published methods for estimating production; they're what I like to call *truth formulas.* In my experience, the numbers they give you are often more accurate than what you may get from some enthusiastic manufacturers.

Predicting wind energy production for a specific home-scale turbine on a specific site isn't something anyone can do affordably with a high level of accuracy, so be conservative in your estimates and round up in your turbine sizing and tower height; then you're likely to be pleasantly surprised. Having all three energy predictions — from the manufacturer and from both truth formulas — is a useful approach. If you take the low number or an average, you'll likely be in the right ballpark.

Formula #1

Hugh Piggott published an article titled "Estimating Wind Energy" in *Home Power* magazine in 2004. In this article, he suggests getting a rough estimate of wind energy production from a given turbine at a given average wind speed with this formula:

Average production (in watts) = (wind speed in mph)^3 × (rotor diameter in feet)^2 ÷ 600

The formula gives you *average* power — not the power at one particular moment — so multiplying by 24 (the number of hours in a day) tells you the daily energy production.

Try this theory out on the Bergey XL.1 that I use in the preceding section. The diameter is 8.2 feet. Take the 8.9 mph average wind speed and plug these numbers into the formula: (8.9 × 8.9 × 8.9 = 705) × (8.2 × 8.2 = 67) ÷ 600 = 79 average watts. Multiply that by 24 hours in a day, and you get 1,896 watt-hours, or about 1.9 kWh per day. On its Web site, Bergey predicts that the XL.1 will produce about 3 kWh per day.

Now try Bergey's larger machine, the Excel, which has a 22-foot diameter. According to Bergey's Web site, in a 9 mph average wind speed, the batteryless grid-tied version of this machine produces about 500 kWh per month, or a bit more than 16 kWh per day. What does Hugh's formula say? (9 × 9 × 9 = 729) × (22 × 22 = 484) ÷ 600 = 588 average watts. Multiply that by 24 hours in a day, and you get 14,113 watts, or almost 14 kWh per day.

Which is right? Well, the proof is in the measured production on your actual site after you've installed the system. Obviously, the formula is an arbitrary and generalized approach, and it doesn't take into account efficiency variations between machines. And manufacturer's numbers have been known to be generous. But the point here isn't to get perfect numbers but numbers within a realistic range. Hugh suggests that numbers derived from his formula will be within 20 percent accuracy.

Formula #2

Mike Klemen of North Dakota has been very involved in unofficial testing of a number of turbines, and he has a reputation for being very detail-oriented and painstaking about testing and reporting. He has a method for checking the claims of manufacturers and inventors, too. Instead of using a formula, he has a table that shows the highest physically possible kilowatt-hours per square foot (using the *Betz limit,* which is the theoretical limit of energy you can extract from the wind) and then a reasonable estimate of the best kilowatt-hours per square foot with typical turbines. You can find these tables and an explanation at www.ndsu.nodak.edu/ndsu/klemen/Perfect_ Turbine.htm.

Applying Mike's approach to Bergey's turbines, you find the following:

- ✔ The XL.1 has 53 square feet of swept area, and according to Mike's second table, you can expect it to generate about 2.7 kWh per day in a 9 mph average wind regime. This is well above the Piggott formula's estimate (1.9 kWh) but below the manufacturer's estimates (3 kWh).

- ✔ The Excel has 415 square feet of swept area, and according to Mike's second table, you can expect it to generate about 21 kWh per day in 9 mph average winds. This is substantially more than either the manufacturer's estimate (12 kWh) or the Piggott formula estimate (15 kWh).

Other Lessons Based on Wind Principles

In addition to the earlier point that energy, not power, is crucial in predicting wind system output, some variation of the following lessons is always high on my list of points to leave with my audience when I do brief presentations on wind energy.

The conclusions in this section are the result of decades of real-world experience. You can likely find people who will tell you that these conclusions aren't correct or that they aren't very important. But please find out where the speakers are coming from before you take their word. Some will be with companies promoting products; others will be newbies who are hopeful and excited but short on knowledge and experience. I come to these lessons from many years of living with wind energy. And in talking with hundreds of wind energy users and dozens of wind energy installers, I find that time and experience leads virtually all of them to similar conclusions.

Dead turbines give no energy

Even when you have energy predictions or even accurately logged real-world production numbers, you don't know everything you need to know about a prospective turbine. Because energy is power multiplied by time, it's worth stating how much time you want to produce this energy — for the long term.

One company may be offering you a great price on a wind turbine that will last only a few months or years on your site. Is that a good buy? Not really. You're purchasing a system to make kilowatt-hours, and that takes time.

To compare prices, you need to consider the cost of the kilowatt-hours over the lifetime of the system. Dividing the total cost by the total lifetime kilowatt-hours gives you a price per kilowatt-hour:

System cost ÷ (predicted yearly kWh × lifetime of system) = price per kWh

Suppose one system you're looking at costs $20,000 and is predicted to make 9,000 kWh per year. If that system lasts 6 years, its lifetime energy cost is 37 cents per kilowatt-hour. Now suppose another system costs $30,000 and is also predicted to make about 9,000 kWh per year. If the second system lasts 12 years, its lifetime energy cost is only 28 cents per kilowatt-hour. You may spend $10,000 more on the second system, but your energy is more than 24 percent cheaper in the long run.

And the calculation considers only the simple cost. Perhaps as important is the aggravation and fear associated with poorly designed and performing wind-electric systems. How many sleepless nights will you endure worrying about whether your turbine will survive? How many dollars and days will you spend replacing or repairing a turbine that couldn't take the conditions you have? I'll always be willing to pay more for more reliability.

Tall towers are essential

If you want to avoid the most common mistake in wind-electric system design, buy and install a tall tower. Across the country and across the world, systems are much too frequently installed with towers that are too short. These short-change the performance of the systems. I strongly encourage you to not follow these poor examples, regardless of whether you're afraid (solution: get good engineering), have restrictive laws (seek a variance), or are concerned about what the neighbors think (get their buy-in).

As soon as you've decided on your specific turbine, tower height is *the* key factor in what sort of production you'll get out of it. And although the standard rule of thumb is to put the lowest blade at least 30 feet above anything within 500 feet, I encourage you to go farther. Think ahead to mature tree height 20 or 30 years in the future, follow the rule, and then round up a bit more. You won't regret it. Why?

Think about the difference in available power between 10 and 12 mph (as I explain earlier in "Wind speed cubed (V^3): A dramatic effect"): The available power at 12 mph is more than 70 percent more than at 10 mph. By going higher with your tower, you can make gains of this magnitude. Think about what happens to the cost per kilowatt-hour and the payback time with this sort of gain, and don't shortchange your system design with a short tower. (Flip to Chapter 14 for the full scoop on the importance of tower height.)

Part II

Assessing Your Situation

The 5th Wave

By Rich Tennant

One of the things we do to conserve energy in the house is to make sure all the lights are turned off before we go to bed.

GOOD TIMES

CARNI

HAPPY DA

In this part . . .

Getting a grip on your situation is the main focus of this part. Chapter 6 discusses your home's energy usage, and Chapter 7 explains how to reduce it through efficiency and conservation measures. Chapter 8 looks at your site's wind-energy potential, because assessing the wind resource is crucial in determining whether wind energy is a viable option for you. Chapter 9 clarifies your relationship to the utility grid. In Chapter 10, I discuss value and payback, and I close this part with Chapter 11, which discusses your options if you choose not to capture the wind.

Chapter 6

The Home Energy Assessment: Gauging Your Energy Appetite

. . . .

In This Chapter

▶ Determining how much energy you want

▶ Exploring techniques for assessing energy

. . . .

*P*eople make budgets or at least have a rough idea of what they need in most areas of their lives. For instance, when you go to the grocery store, you've made an assessment, whether consciously or not, of how much food your family needs for the week. If you buy too little, the kids will go on the warpath; if you buy too much, the fresh stuff will spoil before you have a chance to use it, and you'll have wasted your hard-earned money. The process isn't much different when purchasing a wind-electric system, though people all too often skimp on the energy assessment step. But before you make your own electricity, you need to know how much electrical energy you use and where you use it.

So take the time to do this step and do it well; just follow the guidelines I provide in this chapter. You'll be rewarded with a decent chance of purchasing a system that matches your needs. You'll also be able to identify some of the energy-hungry devices in your house so you can improve your home's efficiency and perhaps buy a smaller — and more affordable — system. (See Chapter 7 for more on saving energy.)

Before You Begin: The Essentials of Energy Assessment

An energy budget works a lot like a financial one. If you're running short on funds, you have a few options: Cut your costs, get a raise, or borrow what you need. When dealing with energy, your choices aren't all that different: You can increase your home's energy efficiency, install a bigger wind-electric system, or get extra energy from somewhere else. Regardless, you should know how much energy you want to use so you can decide how big your energy income should be.

You can safely say that energy uses are on a continuum, from *essential* to *frivolous*. Some things that use energy are critical for life support, and others are for comfort, decoration, or entertainment. I'm not here to judge your energy choices, because everyone has different values. Energy need is subjective, so I just talk about how much energy people want.

In this section, I talk about setting an energy goal based on your wants and discuss why your numbers should be as accurate as possible. You're more likely to make ends meet if you set up your energy budget using real figures.

Setting an energy goal

If you're off-grid, you somehow have to make 100 percent of the energy you want to use and then some — you are the utility, and there's no getting away from providing all the energy you want. If you're on-grid, you can decide to make a little, a lot, or somewhere in between. There are several ways to size a wind-electric system, including the following:

✔ You can do it strictly by energy budget — kilowatt-hours.

✔ You can do it by dollars, how much you want to spend.

✔ In locations with *incentives* (government or utility programs that reduce the financial cost to you; see Chapter 2), you may decide to aim for zero bill — paying nothing for your electricity.

✔ You may be constrained by your site, regulations, or existing equipment.

Usually, though, your best bet is to focus on your energy wants first. The two fundamental numbers you need for wind-electric system design are

✔ Your energy wants in kilowatt-hours (kWh)

✔ Your wind resource in average miles per hour (see Chapter 8 for details)

I recommend that you start with *zero* outside energy as your goal. That is, start by designing a system that will make all the electricity you want over a year's time. If your bank account can't afford it when you're shopping for your system, you can always back off and design a smaller system. But zero energy is a good initial target, and it makes the calculation simple, because your energy generation goal equals your energy use. (You find details on how to calculate your energy use later in this chapter.)

This system sizing process sometimes works the other way around, especially with off-grid systems: Instead of figuring out how much you want to generate, you need to figure out what you can afford to generate and then see whether you can live with it.

An investment in renewable energy doesn't have to be all or nothing. You may decide to buy a system that makes only half of your electricity. If you're on-grid, the utility picks up the rest and bills you for it. If you're off-grid, you have to make the rest of your electricity from other sources. See Chapter 11 for renewable energy sources you can use instead of wind or combined with it to meet your goals.

The importance of good numbers: Quantifying your wants

A friend and colleague in the renewable energy industry likes to say that people don't want electricity — they want lights, TV, and cold beer. His point is well taken. When I ask clients how much energy they need, they rarely answer in electrical units. They may say, "Well, we want lights, a stereo, a fridge, and a computer for our cabin in the mountains and all the normal appliances in our house in the city."

Thinking more specifically about the energy you want is useful. If you tell me you want "lots of electricity," I can only tell you it will cost "lots of dollars." But when you start listing the actual appliances you want, you're getting a good start on energy assessment. The next step is to convert the description of what you want to power into numbers.

I'm surprised how regularly people make important financial decisions based on guesswork. For example, what happens when you mail-order clothing by saying, "Well, I think she wears a size 12" or "My recollection is that his sneakers are size 9"? You usually won't get very good results. You want to confirm important data before wasting your effort and dollars. Confirmation is even more important when you're installing an energy system.

You need to know how many kilowatt-hours you want each day. If you can cut to the chase and get this number with less pain, go for it. But make sure it's a good number, not a wild guess. Without a solid energy number, your design — and therefore your project budget — will be questionable. Trying to pin down how much energy you want isn't glamorous work, but without a good energy estimate, you have three possible outcomes:

- ✔ You get very lucky and design a system that makes just the amount of energy you need. (This is extremely unlikely.)

- ✔ You purchase a system that doesn't make as much energy as you want, and upgrading may be very difficult and costly.

- ✔ You spend more money than necessary to purchase and install a system that makes more energy than you need.

Do you speak kilowatt-hours?

You know gallons of gas. You know ounces of coffee. You know pounds of ground beef. I think American culture isn't focused on energy primarily because energy has been so cheap and abundant. As you face the reality of a limited energy supply, however, you need to start thinking energy.

To flourish in an economy of limited energy and to make the most of renewable energy systems, you need to understand, think, feel, speak, and live kilowatt-hours. Kilowatt-hours need to become as familiar to you as your own shoes. Train yourself to notice the nuances of energy use, reduction, and production. This can help you reduce your energy dependence and become the energy manager you need to be if you want to make your own electricity.

Make sure you confirm important data and invest the time, effort, and even money that you need to do an excellent job at this stage of your design process. You won't be able to replace an energy system the way you can exchange a new blouse or pair of shoes. In the next section, I explain everything you need to do to calculate your energy numbers.

Examining Energy Assessment Methods

In this section, I tell you how to determine how much electrical energy you want, because wind-electric systems primarily address electrical generation, use, and efficiency. In Chapter 7, I talk about energy efficiency and the bigger picture, including thermal efficiency and other fuels besides electricity.

Here are the goals of an electrical energy assessment:

- ✔ To understand how many kilowatt-hours the existing home uses or a new home will use

- ✔ To understand where you're using these kilowatt-hours

- ✔ To help you understand where you can reduce the usage (see Chapter 7 for info on making your home more efficient)

You can take three basic approaches to an energy assessment:

- ✔ **Detailed analysis:** I recommend this for all homes if possible. It gives the most information and therefore leads to the most efficient homes and the most accurate wind-electric system design.

- ✔ **Overview based on past usage:** With on-grid homes, this can be an accurate big-picture view (from the utility bill), but it isn't detailed.

> ✔ **Estimating, guessing, and wild guessing:** This, unfortunately, is a very common approach, but it's a frustrating one for designers and those who must live with the consequences of "guess designing."

Which approach you use depends on your situation (specifically, whether you have an existing on-grid home, an existing off-grid home, or a new home), your budget for analysis, and your perseverance. Remember that you get what you pay for — if you want a high quality design, give it the best chance by doing a high-quality energy assessment.

One important consideration while you're doing a load analysis is whether your usage will change — what changes are coming up in your lifestyle? Will new children or grandchildren join the family? Are the kids heading off to college soon? Are you considering adding a woodworking shop or pottery studio? Will one of you retire soon and spend more time at home? All these possibilities and more can affect your future energy use. If you expect significant changes in your energy use in the future, you should measure or estimate the specific loads and adjust your system design accordingly. In most cases, energy use will grow (unless you have teenage kids about to leave for college), and it's best to round up somewhat to cover this growth.

On-grid homes: Using utility bills and a detailed load analysis

On-grid folks have it easy. If you're a poor record keeper, you can lean on the utility and get their employees to send your energy history. You know they keep track — that's how they make their living. But I hope you dig deeper and get the detail. The more you know, the wiser you can be with your energy use and generation. In this section, I discuss both interpreting utility bills and performing a more detailed analysis.

Starting with utility bills: How much energy you use

If you have an existing home serviced by a utility, the big-picture part of your energy assessment job is a snap. Just gather up the last year's electricity bills (see a sample in Figure 6-1), tally them all up, and calculate average kilowatt-hours (kWh) per day. In other words, figure out how many kilowatt-hours you've used for the year and divide by 365.

Some utilities bill on a monthly cycle, and others bill every other month. A year's history of electricity bills gives you a clear and accurate overview of the home's electricity usage throughout the seasons. Some utilities will even mail, fax, or e-mail you a summary of the past year's history.

All utility bills seem to complicate the numbers with multiple billing categories (refer to Figure 6-1). Take a careful look at two numbers to make it easy:

✔ The total kilowatt-hours you used

✔ The total money the utility charged

The details of exactly what's charged for what are unimportant, but contact your utility if you have any questions about how many kilowatt-hours you're using. Divide the total money charged by the total kWh used to get your cost per kWh.

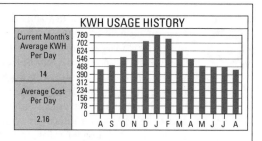

Figure 6-1:
A utility bill is a helpful tool in assessing energy usage.

Easy Street Electric Cooperative
P.O. Box 1
Anywhere, USA 00000-0000

Office hours:
7:30 A.M. to 4:30 P.M. Monday–Friday
800-555-1234
www.easystreetelectric.com

JOHN A. DOE
123 MAIN STREET
ANYWHERE, USA 00000-0000

KWH USAGE HISTORY	
Current Month's Average KWH Per Day	14
Average Cost Per Day	2.16

SERVICE SUMMARY

Account Number:	9999999999
Name:	JOHN A. DOE
Location Number:	99A999999
Billing Date:	09/24/2009
Due Date:	10/14/2009
Service Dates:	08/01/2009-09/01/2009
Days of Service:	31
Type of Bill:	REGULAR

ACCOUNT ACTIVITY

PREVIOUS BALANCE	99.99
PAYMENTS	-99.99
ADJUSTMENTS	0.00
BALANCE FORWARD	0.00

FACILITY CHARGE	20.00
ELECTRIC SERVICE	33.11
SECURITY LIGHT	9.50
SALES TAX	3.50
WHOLESALE POWER COST ADJUSTMENT	0.95
PUBLIC BENEFITS PROG	1.33

READINGS THIS PERIOD

Meter Number:	088888888
Current Reading:	99990
Prior Reading:	99560
Multiplier:	1.0000
KWHs Used	**430**
Rate:	01

Due Date	10/14/2009	Net Due	68.39
Gross Due After	10/14/2009	Gross Due	69.41

Giving electric heating the cool treatment

Heating is a huge energy demand that you can often meet by means more economical than electrical resistance heat, which is a dirty energy gobbler, considering that most electricity in North America is made with coal and oil and must be transmitted to your home. (There is one form of electric heat that is much more efficient — heat pumps. These can be a very good match with wind-electric systems.) American homes that don't heat with electricity typically use 25 to 30 kWh per day (Europeans use much less). If you heat or air-condition with electricity, that number may be three to four times as much. If you live simply and efficiently, the number may be 15 kWh per day or less. And if you use propane or natural gas wherever possible, it may be lower.

I recently did an energy survey on a house that used only 6 kWh of electricity per day. This modern home had careful occupants but no special or expensive energy efficiency measures. One major factor was that all appliances that could run on natural gas did — water heater, range, dryer, furnace. I'm not necessarily advocating this — I prefer renewable sources — but it does give you a clue that heating a home uses a lot of energy. You need to examine your motivations and decide what's most important to you. If saving dollars is your highest goal, you have a tough choice, because the costs of electricity and fossil fuels vary a lot and change over time. If using energy you generate on site is your goal, look hard at renewable ways to heat, such as passive solar design, wood, and heat pumps.

This book focuses on electricity, so I don't go into much detail on heating choices. However, because some homes do have electric heat and most use some form of heating, I include some thermal efficiency advice in Chapter 7. Check it out. Even if thermal efficiency won't affect your wind-generator size, it's still a worthwhile investment — for both you and the environment.

Moving to a detailed load analysis: Where it all goes

If you're able to get an accurate history and calculate an average kilowatt-hours per day from a year's worth of utility bills, you're off to a good start. Some people stop here, but I encourage you to take it further and identify where you can save energy.

If you've determined that your on-grid home uses, say 28 kWh per day average, the next step is to get at least some idea of where you're using all that electricity. Figure 6-2 shows a typical distribution of electricity (not total energy) in a typically wasteful American home that isn't primarily heated with electricity. You can see that space heating and cooling, water heating, and lighting are the big ones. Refrigeration and electronics are serious users, too, and the rest adds up.

So you can take your 28 kWh and divvy it up according to those percentages, but this will be accurate only if you're an average American. Personally, I don't know any average people — the way people use energy varies, depending on their values, lifestyle, and so on.

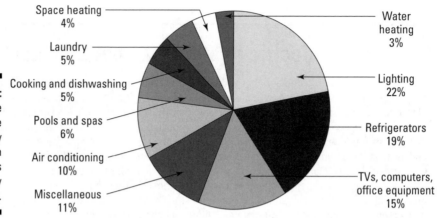

Figure 6-2:
An example
of home
electricity
distribution
(heat is
primarily
nonelectric).

Space heating 4%
Water heating 3%
Laundry 5%
Lighting 22%
Cooking and dishwashing 5%
Pools and spas 6%
Refrigerators 19%
Air conditioning 10%
TVs, computers, office equipment 15%
Miscellaneous 11%

To figure out where your electricity goes, make up a list or spreadsheet cataloging all the electrical loads in your home (ideally detailing each light and each appliance — fridge, washing machine, range, and so on). Give yourself columns for the name of the appliance (or other load), its draw in watts, and the total watt-hours per day. Then start filling in the blanks as much as possible. Here are several methods for getting these numbers, starting with the best and ending with the least accurate:

- ✔ Measure as many loads as possible using a plug-in watt/watt-hour meter left on for several days to get realistic measurements. This tool (see Figure 6-3) allows you to measure real-world usage of specific appliances, especially cycling appliances such as refrigerators and freezers. Chapter 4 explains more about wattmeters.

- ✔ Measure wattage of loads with a wattmeter under typical running conditions and multiply by the estimated or timed usage per day.

- ✔ Calculate watt-hours using the rated wattage of appliances and the estimated hours of use. Nameplate information, which appears on most appliances, is useful for this task.

- ✔ Use standardized estimates of typical American energy use per appliance, or use Energy Star numbers for your specific appliances. The Energy Star Web site (www.energystar.gov) and the American Council for an Energy Efficient Economy (www.aceee.org) are both good sources for generalized numbers.

You can use your utility meter and your home's circuit breakers to isolate and measure individual circuits or major appliances. This requires knowing how to read your meter (very complex in some cases!) and being willing to shut down everything but one circuit for periods of time. But if you're up for it, this can be an accurate way to measure devices such as your water heater, heat pump or furnace, dryer, and so on. Or you may even be able to buy an aftermarket whole-house energy meter that monitors and logs each circuit.

In the real world, you'll likely end up using a combination of these methods, because directly measuring some loads — such as 240 VAC loads and hard-wired loads like your water heater or dryer — is impractical or impossible. But don't let this slow you from doing all you can by real-world measurements.

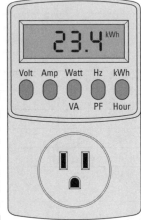

Figure 6-3:
A watt/watt-hour meter measures specific electrical loads.

Plug all your measurements into a spreadsheet, and — presto! — you have a detailed load analysis. Figure 6-4 shows a sample load analysis, with most of the appliances and other loads you'll find in a typical home. This sort of detail is ideal — get it down to the lights in each room and all the appliances.

Try it out for your home. Go through the chart and fill in as much as you can. For plugged-in loads that you switch on and off, either measure the wattage using a watt meter or look at the name plate to get the max wattage. Put in the hours you estimate (or time) that you use each load each day. Multiply watts times hours to get watt-hours.

For cycling appliances like your fridge, furnace blower, or maybe even your computer system, it's best to measure over time with a watt-hour meter. On these appliances, you don't need to fill in watts and hours because you're cutting right to the chase and measuring watt-hours.

A detailed load analysis can be eye-opening. It may steer you to very low-hanging fruit — areas ripe for improvement. For instance, if your refrigerator uses more than 2 kWh per day, you can replace it with one that uses less than 1 kWh per day while giving you the same functionality. Such energy efficiency and conservation measures have the potential of cutting your energy requirements by 10 to 50 percent. Chapter 7 has more tips on how to do this in your home. You'll want to adjust your spreadsheet (down I hope!) as you reduce energy usage through efficiency measures. As soon as you have a solid energy figure, you have one of the key pieces of information to start working on sizing your wind-electric system (see Chapter 16).

Appliances	Number	Wattage	Hours per day	Watt-hours per day
Kitchen				
Refrigerator				
Coffee maker				
Toaster				
Microwave				
Freezer				
Stove/oven				
Living Room				
Television				
DVD player				
Stereo (include speakers)				
Lamps				
Laundry Room				
Washing machine				
Dryer				
Bathroom				
Hairdryer				
Electric shaver				
Electric toothbrush				
Bedrooms				
Clock radio				
Television				
Bedside lamps				

Figure 6-4:
A sample
load
analysis of
an on-grid
home.

Appliances	Number	Wattage	Hours per day	Watt-hours per day
Office				
Computer				
Printer				
Fax machine				
Telephone				
Answering machine				
Scanner				
Modem				
Cell phone chargers				
Desk lamps				
Other Appliances				
Air conditioner				
Vacuum cleaner				
Fans				
Chest freezer				
Appliance Total				

Figure 6-4
(continued)

Lights	Number	Wattage	Hours per day	Watt-hours per day
Living room				
Kitchen				
Bedroom 1				
Bedroom 2				
Bedroom 3				
Bedroom 4				
Bathroom				
Laundry room				
Hall				
Office				
Outside				
Lights Total				

Figure 6-4 (continued)

Appliance Total		
Lights Total		
TOTAL DAILY ENERGY USAGE		

Off-grid homes: Extrapolating from current energy sources

Off-grid homes present a load-analysis challenge in some ways and can make the process easier in others. They're a challenge because there's typically no single meter that measures all energy used, as with on-grid homes. They can be easier because the load is often smaller and easier to quantify and because the users are often more conscious of energy use.

To measure the total load in an off-grid home, one simple method is to install a utility style meter on the inverter's output. But very often, off-grid design happens before the house is built or lived in, so you're back to the spreadsheet. (See Chapter 3 for info on inverters and other system components.)

Doing a complete load analysis (like the one in the preceding section) is still ideal in this case — it gives you the best possible information to design your wind-electric system. You may have both AC and DC loads, and you need to include a factor for battery inefficiency and charging (see Chapter 15 for more on this). The complete load analysis can uncover all these details.

But even without an electric bill, you may be able to find some shortcuts using the existing electricity supply of an off-grid home. Off-grid homes typically have more than one source of electricity. Common sources include solar electricity and a backup gas generator (and if you're blessed with falling water, hydroelectricity). If you're lucky, these have amp-hour or watt-hour metering logging the production of the photovoltaic (PV) array and the generator. The typical goal in adding a wind generator is to reduce gas generator run time, so looking at how much energy the gas generator is making can give you an idea of what your desired wind production is.

Here are a couple of shortcuts for doing a load analysis on different off-grid electricity sources if you have no metering:

- ✔ **Make a rough calculation of your solar-electric production by looking at the rated capacity of your array and the *sun-hours* (solar energy available) in your area.** For example, imagine you have a 1,600 W solar-electric array and your area gets 4.5 sun-hours. A simple formula multiplies these two numbers and then applies a factor of 0.65 to get actual watt-hours produced. So in this case, $1,600 \times 4.5 \times 0.65 = 4,680$ watt-hours per day, or about 4.7 kWh.

- ✔ **Guess how much electricity you're using from your gas generator by considering its rating and the hours of generator run time.** This number will likely be a high estimate, because gas generators rarely run fully loaded. Suppose you have a 3,000-watt gas generator and you run it 16 hours per week. Theoretically, this would mean an additional 48 kWh per week ($3 \text{ kW} \times 16$ hours), or about 7 kWh per day. In most cases, you're probably using only about half that much, depending on your setup. But using the high number in your design may be safer.

You can see that the preceding methods are pretty rough. Again, doing a detailed analysis of the loads is much better. Whichever method or combination of methods you use, these tools can get you some sort of estimate of how much energy you're using and how much you want your wind-electric system to make.

New homes: Making a detailed load list and reviewing past home history

If you're building a new home, your situation presents special challenges. You don't have a utility meter and history. You don't have an existing renewable energy supply with metering. And you may not even know just which loads you'll use in the house.

Here again, a detailed load analysis (like the one in Figure 6-4) is your best friend. In this case, you need to make a list of all the appliances you *intend* to use, not the ones you're already using. Take the time and do the work to map out each load (or set of loads) and make an intelligent estimate of the run time and total watt-hours.

One other tool may help you: your past energy use. Although the house itself has some impact on energy use, the occupants have a greater impact. So looking at records from your existing home is a worthwhile pursuit. Sometimes you may even have records from two homes, a city home and a vacation home. Extrapolating from all the information you have can give you the best number available.

Imagine, for example, that you use 32 kWh per day in your city home, and in the weekend home you're selling, you used 18 kWh per day. You can look at the loads and time you spent in each and make some educated guesses about what you expect usage in your new home to be. This depends on the specific loads you use and how much time you spend in the home. However, the truth will come only with real experience with your new wind-electric system.

Chapter 7

Increasing Your Home's Energy Efficiency

*B*efore you go investing big bucks in a small-scale wind system (or any other renewable energy source), you need to supercharge your efficiency efforts. Energy efficiency is the best investment you can make — financially and environmentally. Amory Lovins, who is the cofounder, chairman, and chief scientist of the Rocky Mountain Institute, an energy think tank, coined the term *negawatts* to mean the energy you don't have to generate due to energy efficiency strategies.

Getting excited about putting that spinning wind turbine up is easy, but doing the energy efficiency work is vital. The results of this work can make a wind-electric system much more affordable. If you can cut down on the size of the system you need, you cut down on the cost as well as on the environmental impact of your home and your system. Remember — the only free energy is energy that you don't have to make.

In this chapter, I show you how to start down the path to greater energy efficiency by figuring out your current costs and establishing your goals. I then explain how to increase both thermal and electrical efficiency and get rid of phantom loads, no matter what the type of device. I wrap up by walking you through the process of calculating how much energy you've saved. This is perhaps the most important chapter in the book, because energy efficiency work is the most important work you can do. My coverage here is only an introduction, and I recommend that you seek out more information through books (such as *Energy Efficient Homes For Dummies,* by Rik DeGunther, or *The Home Energy Diet*, by Paul Scheckel), Web sites (`aceee.org` and many more), workshops, and the direct experience of others.

Taking the First Steps to Greater Energy Efficiency

Do you like traveling without a destination, with no idea of the costs? This sounds like fun — once in a while — but it's not a good way to approach most renewable energy projects. When beginning to add renewable energy to your home, you need to take a few steps before making any changes to your home's energy use: tallying current costs and setting a goal for energy reduction. This section walks you through the process of calculating current usage, finding ways to reduce energy use, and setting a goal for energy reduction both before you install your renewable energy system and after.

Tallying the shocking current system cost

Some folks' excitement for renewable energy leads them to get a dealer out on their property to scope out the options. This typically results in a rough quote — and a reality check on the amount of energy they're using.

Here's an example: Suppose I go out to a home and do a brief energy analysis using Terry's utility bill as my source (see Chapter 6 for details on this method). I conclude that she uses 36 kWh per day, a bit more than the national average for nonelectrically heated homes in the United States. I do a cursory site analysis and see that the wind maps show a modest wind resource. I specify that a 140-foot freestanding tower is necessary and a machine with about a 30-foot rotor will likely make a bit more energy than she needs over the course of the year.

Now we get to the numbers on how much a complete system will cost. At this stage, they're a rough estimate, but they're large. I tell Terry that the installed cost of a wind-electric system to make all the energy for her home will be about $90,000 minimum.

Terry is shocked. She had no idea that it would be this much and is discouraged about the prospects of actually installing and using renewable energy. Now she has several options. She can give up. She can go visit her banker or relatives. She can consider other renewable energy options (like the ones in Chapter 11). She can also consider a wind-electric system that won't produce all her energy.

Although all these options are on the table, the best option for anyone interested in a wind system is to do the hard work of identifying the home's energy load and reducing it. Why am I so adamant in recommending this? Energy efficiency is almost always less expensive than generating electricity. Identifying and reducing the home's energy load often significantly reduces the size of wind-electric system recommended and in most cases leads to reduced cost for equipment and installation.

Need proof? Take a deeper look at Terry's example. Although the utility bill gave me the big picture — 36 kWh per day — it doesn't tell me anything about where she's using all that energy. To find out where all the energy is going, I have to closely examine the electrical energy use in her house (see Chapter 6 for the steps to follow in your home). Here's what I find:

- ✔ A fridge that's on its last legs uses 3.75 kWh per day
- ✔ A poorly insulated electric water heater set to 140 degrees
- ✔ All incandescent lighting
- ✔ Outside lighting that's typically left on all night
- ✔ 2.5 kWh of *phantom loads* (electrical loads that continue to use energy when they're switched off) — all relatively easy to eliminate, as I explain later in this chapter

My recommendations to Terry and the kWh savings are as follows:

- ✔ Replace the fridge with a super-efficient 0.75 kWh per day model: 3 kWh saved
- ✔ Set the electric water heater to 120 degrees, insulate it, and put it on a timer: 5 kWh saved
- ✔ Replace incandescent lighting with fluorescent lights and LEDs: 4 kWh saved
- ✔ Put outside lighting on sensors: 2 kWh saved
- ✔ Eliminate phantom loads with power strips and switches: 2.5 kWh saved

Terry's energy efficiency moves cost her about $2,000 and save her 16.5 kWh per day, cutting her usage by 46 percent and making her renewable energy system design very different. (This calculation doesn't even count any moves in thermal efficiency she eventually can make. These will save her energy — if not electricity — and will be dollars well spent as well.) Now she needs only about a 25-foot rotor (rather than a 30-foot rotor), and the total installed cost of a wind-electric system would be roughly $60,000. (The savings aren't linear, because tower cost is a major portion of a wind-electric system, especially where trees are tall).

Sixty grand is still a chunk of change, so Terry is ruminating about whether she can raise the cash. But $60,000 is more attainable than $90,000. And meanwhile, Terry saves about $600 per year for her $2,000 investment, a very good return and a far cry from the modest return she'll get if she decides to go ahead and invest in a wind-electric system.

Your goal: Cutting 20 to 60 percent

What's a reasonable goal for increasing energy efficiency? An average American should be able to cut his or her home energy use by 20 percent. Cutting 20 percent of home energy use doesn't take much thinking, looking, working, or spending. If your house is particularly wasteful or you're particularly energetic, you can aim for much higher reductions, even exceeding 60 percent of your present energy use.

And why not? Reducing your energy use isn't glamorous work, but are you interested in renewable energy for reasons of glamour or to save money and the planet? No matter what your motivations are, I bet saving money is attractive to you. (If not, send me the extra cash you have lying around.)

Generalizing about your energy-saving goal is difficult. Some of you may already live in very efficient homes, where reducing your usage by 60 percent may be nearly impossible (without turning off the fridge). Others may live in very wasteful homes, where attaining this goal may be reasonable.

I suggest that you start with a goal of 20 percent reduction. If you find that really easy, go for another 20 percent. If it's hard, either you've already done the work, or you're letting yourself off easy for being an energy hog.

Read on to find out what to look for and how to modify your home so you can save money on your energy bills. Reducing energy use will shrink the size of the wind-electric system you need and get you closer to your renewable energy dreams.

Boosting Your Thermal Efficiency

A large number of homes in North America don't use electricity for space heating. Many people use natural gas, propane, fuel oil, coal, wood, and other fuels. Other homes do use electricity, which means they're indirectly using coal, oil, gas, hydroelectricity, and other minor sources to heat their homes (and they are using much more of these sources of energy than they would if they were to use them directly). Home cooling is typically electrically sourced.

Space heating and cooling almost always make up the largest energy load in a typical American home. This means that thermal efficiency is a key place to focus, regardless of how you heat and cool. In this section, I describe several areas of thermal systems in your home to evaluate and change for greater efficiency.

If you want to get really serious about increasing your home's thermal efficiency, hire a pro to come in with a blower door (see Figure 7-1) and a thermal imaging camera to get down to the nitty-gritty of your home's thermal performance. These two tools and others can isolate the most serious culprits in your house and point you toward the most cost-effective solutions.

Considering heating and cooling sources

Because home heating and cooling (depending on where you live) are typically the largest single energy load, it's in your best interest to heat and cool as efficiently as possible, regardless of the energy source. This section covers some of your options. Local heating and cooling contractors are one source of information about the state-of-the-art in efficient systems. You can also seek out more-impartial advice from organizations such as the American Council for an Energy-Efficient Economy (aceee.org).

Regardless of your heating or cooling source, your system should use a programmable thermostat (if your system has a thermostat at all). That way, you can make sure you minimize heating and cooling efforts when you're away at work or asleep for the night.

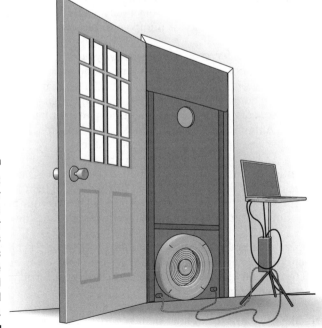

Figure 7-1:
A blower door can test your home's leakiness and guide you toward thermal efficiency.

Heating sources

Whether for new construction or retrofit, deciding on the heating fuel to use is difficult, especially if cost is your only consideration. Fuel costs vary over time, sometimes dramatically. Choosing an oil- or natural gas-fired furnace today because of present prices may come back to haunt you later.

Other factors may have a big influence on your fuel choice as well. For example, people who have environmental values may lean toward passive solar design, wood heat, or more conventionally, electricity. If you don't have the right situation for passive solar design or wood heating, I favor electricity for the following reasons:

- Coal, oil, nuclear, and gas will always be nonrenewable, so I tend to avoid them as fuels.

- In some parts of the country, electrical supply is partially renewable already (40 percent hydroelectricity in my neighborhood).

- With electricity, you have the option of making your own with renewable sources or of buying *green power* from your utility (see Chapter 11 for info on applying for green power).

Other people base their choice of heating fuel sources on reliability. Although electricity may be greener, you certainly can't back up an electric heating system with a battery bank — it's just too big of a load. If you need outage protection, having a tank full of fuel is very reassuring (though you still may need battery or generator backup for pumps and blowers).

Whatever fuel you choose, focus on a high-efficiency heating system, recognizing that spending more money upfront often pays off down the road. Talk with at least two local heating contractors, and try to find ones that your local renewable energy experts recommend. Think simplicity — too often, heating contractors get you into very complex systems. And make sure your heating system is well maintained. Although most homeowners can clean filters and registers, having the pros check equipment annually is worth the price you pay.

One sensible option not used often enough in green homes is heat pumps. Heat pumps are not actually *creators* of heat but *movers* of heat. The pumps gather heat from the air, ground, or water and move it into your house, or for cooling, they pump heat from your house into the air, ground, or water.

Although people talk about fuel-sourced heating systems on a scale of 100 percent efficiency (with the best being in the 90s), on that scale, heat pumps are about 200 to 500 percent efficient: They use 1 kWh of electricity to move 2 to 5 kWh of heat from their source.

I particularly like the marriage of heat pumps with radiant floors, a combination that results in a comfortable, healthy, and energy-efficient heat delivery method. A *radiant floor* has pipes with heated water running right under your feet — or your belly as you relax on the floor — a pleasant, efficient way to get heat to you.

Discuss the following heat-pump options with your contractor to get local advice and comparative performance and pricing:

- ✔ Ground-source heat pumps have the highest efficiencies, but they're also most costly to install. (These pumps are also sometimes called *geothermal,* not to be confused with hot springs.)

- ✔ Air-source heat pumps are the least costly, but they have the lowest performance. Often these are the best overall choice.

- ✔ Water-source heat pumps are less common, using drilled water wells or banks of pipe in a pond. Most people do not have this option readily available, but they can be the highest performance of all.

Cooling sources

With cooling systems, you generally don't have a choice of fuels — most methods of artificial home cooling are electrically powered. Again, focusing on the energy efficiency of your air conditioning system will reap benefits, and heat pumps can be an excellent option if you have a forced air heating/ cooling system. Your friendly local HVAC contractor is your best resource here, but arm yourself by doing more research on the common systems installed in your area before signing up for the sales pitch. Finding renewable energy experts in your area can steer you toward contractors who are focused on clean, green, and efficient systems.

Looking at the design and landscaping of your home can often decrease heat gain through shading, natural ventilation, and materials choices. Here's how:

- ✔ Design your home with overhangs that protect the house from the hot summer sun.

- ✔ Use plantings to provide shading and moisture around your home.

- ✔ Discuss options for protective glazing with local window contractors.

- ✔ Design or remodel your house to increase natural ventilation, using the tendency for hot air to rise as well as natural winds. Consider locations of windows and doors to take the best advantage of natural cooling.

Passive solar designers in your area are great resources for information on the best cooling strategies, taking into account climate, humidity, available materials, and so on. For more information on passive solar heating, check out *Solar Power Your Home For Dummies,* by Rik DeGunther (Wiley).

Looking at heating and cooling ducts

How often have you seen uninsulated and unsealed ductwork exposed in attics, basements, crawl spaces, or even outside homes? This is a major source of energy loss. Air leakage from forced air heating and cooling ducts can commonly run to 15 or 20 percent of the air or more.

In addition to installing ducts carefully and insulating them, sealing them is crucial to good performance. Foil/metal tape made especially for duct sealing is one basic tool, and sealing mastic, which comes in a bucket, is another. Heating and cooling pros can test your ductwork for leaks and take appropriate measures. You should do this *now* and then again every 8 to 10 years. If you're heating or cooling the great outdoors, seal things up! You may reduce your energy bills significantly.

Evaluating the insulation

In terms of thermal energy efficiency, you want heat, but you want it to stay where it's needed. Proper insulation is a key part of thermal energy efficiency. Insulation comes in a variety of forms, and each type has advantages and disadvantages. I encourage you to choose the more benign forms and installation methods, such as blown-in cellulose (shredded newspaper) and other natural products. Local insulation contractors can counsel you on what's available and at what cost.

How much insulation should you use? My advice is more than you think you need. Find out what the standard recommendations are for your area and round up. This is infrastructure for the long haul, and if the insulation is installed well, it'll save energy, money, and the environment for many decades. A super-insulated house is your best defense against rising energy costs, and you can lower your environmental footprint and increase your comfort in the bargain.

Proper installation is as important as having the right product. Poorly fitted insulation does a poor job indeed. The goal is to trap air and keep hot air from escaping. Regardless of the kind of insulation, meeting this goal is tough if you leave large gaps in the insulation.

Wrapping up with the building envelope

Closely associated with insulation is the tightness of your building envelope. Without a tight building, lots of insulation may not give you the gain you hoped for. Moving air moves warm and cold air around, so poor sealing around doors, windows, outlets, and other building openings cost you money and energy for the life of the building. I'm acutely aware of these issues, because I live in a leaky house, which I'm gradually trying to seal up.

Strategies for a tight building envelope include the following:

- Wall systems that don't include studs or other *thermal breaks* (systems such as structural insulated panels [SIPs] and Rastra [recycled Styrofoam and concrete]); not only do these systems have superior insulation, but they also provide less opportunity for air infiltration

- House-wrap around the sheeting of a house that cuts down on air movement while allowing moisture movement

- Caulking around all windows and doors and other breaks in envelope that allow leakage

- Vapor barrier (special plastic sheeting that limits moisture transfer) and insulation around the perimeter of the foundation

Finding a good building envelope mentor or contractor can be challenging. Again, I recommend starting in the renewable energy and passive solar design fields, finding the local professionals. These people may not be able to help you directly, but they will know who is doing this work and who is doing it well.

Another straightforward strategy is to find the company or consultant who owns and uses a *blower door*. This device pressurizes your house and quantifies the air loss. People who own and use these not-that-inexpensive devices usually have training in using them and in fixing the problems they find.

Some sealing contractors charge based on how much improvement you see. My wife and I are having our building envelope tightened this year. They blower-door test before and after, and we pay only per percentage of infiltration reduced. Search for someone who does blower-door testing, and you'll likely find someone who can do or subcontract the sealing.

If your house is very well sealed, you need to make sure it has enough air transfer so the air quality in your home doesn't suffer. And if you add a ventilation system, you may consider a *heat recovery ventilator* (HRV), which grabs some of the heat in the air you're venting out so you don't lose it. Both of these topics require expertise that I and most homeowners don't have, so get some local help from a ventilation nerd.

Increasing Your Electrical Efficiency

The efficiency of your electrical appliances directly affects the planning, design, and cost effectiveness of a wind-electric system. Every watt-hour you don't consume is a watt-hour you don't need to generate, transport, store, convert, and monitor. Electrical energy efficiency is often easier to measure than thermal energy efficiency. In many cases, you can do direct measurements on specific appliances with readily available metering. Your efforts will also show up on the bottom line of your electricity bill. So you can lower your bill, decrease your environmental footprint, and quantify your progress at the same time.

In this section, I describe a variety of devices you can change to boost your home's electrical efficiency.

The complexity of life is directly reflected in your energy appetite and therefore in your energy bills and environmental footprint. Simplifying your life by decreasing the size of your house, questioning energy-using appliances, and looking for ways to do tasks unplugged can shrink both your bills and your footprint, sometimes radically. For some tips on cutting out the clutter, flip to Chapter 11.

Refrigeration

When people design off-grid homes, any significant heating loads are shifted to other fuels — sunshine, wood, or fossil fuels. This usually means that the largest single load in the home is the refrigerator. This appliance is essential in most American homes, and it's not one that you can turn off when you're not using it or when there's not enough energy.

Because the refrigerator is in constant use, getting an efficient fridge is extremely important. A recent energy survey at a local resort found one fridge that used only 0.6 kWh (600 watt-hours) per day and another similarly sized fridge that used 4 kWh per day! This is a case where spending more upfront pays off over the long term.

For instance, in 1984, my wife and I paid $2,700 for our super-high-efficiency refrigerator. This is an almost custom machine designed for the off-grid market, and it's a real energy sipper. Even to this day, it's probably the most energy-efficient upright fridge on the market. And that was very big money for us in those early days on our homestead. But if we had waltzed into town and bought a cheap fridge at the local appliance store for a few hundred dollars, it would've been very expensive for us in the long run. Because we're off-grid, we would've had to purchase thousands and thousands more dollars of solar and wind-electric capacity or run our propane generator many hours, an option that would've been expensive and dirty as well.

Often I recommend not spending that much on a fridge for an on-grid home because some good options are available in mass-market refrigerators. The renewable energy–specific products are still more efficient, but they tend to be direct current (DC), not as modern, and not as full-featured as the best mass-market products, which are often only 15 or 20 percent less efficient.

Investing well in a refrigerator is very important. If your present fridge uses more than 2 kWh per day, I recommend replacing it. A modern fridge that uses less than 1 kWh per day is relatively easy to find and afford. To help you decide whether to invest in a more energy-efficient model, plug your fridge into a watt-hour meter for a typical week to find out what it's costing you.

As for separate freezers, ask yourself whether you really need more than the freezer in your fridge, and buy the smallest freezer you really need. If you're storing frozen food for more than 6 months, reconsider your plan. Freezers are high-energy devices — a small one with food cycled more frequently is much better than two monsters. Put your freezer in a cool environment, such as a garage or basement, so it's not working overtime to counteract your home's heating system.

For both fridges and freezers, setting the thermostats up as high as possible and still getting the cooling results you desire saves energy and money. Cleaning the coils on the back of the machine and giving enough air space for the heat to exit helps, too. And so does not standing there in your jammies with the door open!

Hot water

Heating domestic water is very often the largest electricity user in an on-grid home (if another energy source is used for heating and cooling). Reducing this energy load reduces your overall load and, eventually, your wind-electric system cost and size. Strategies to reduce your dependence on electricity or gas for heating water include the following:

- ✔ Installing a solar hot-water system; this investment may be in the $8,000 to $16,000 range, but it can often carry 75 percent of your hot water load

- ✔ Super-insulating your water heater tank

- ✔ Turning the thermostat down to 120°F

- ✔ Putting your electric water heater on a timer; if you always shower in the morning, you may be able to have your heater run only an hour a day

- ✔ Considering a *demand heater*, which heats water only when you turn on the faucet (***Note:*** Demand heaters are best if your usage is modest. If you have only a few people in the house and you don't shower three times a day, an on-demand heater may save you energy. If you're water hogs, an on-demand heater opens up the door to endless showers and endless energy use to heat the water.)

Lighting

Lighting is the low-hanging fruit of electrical energy efficiency. Almost everyone, no matter what budget or ability, can implement changes suggested in the following sections. Most of the products I recommend are commonly available even in small-town America. In the end, there's really no good excuse for using inefficient lighting.

Using efficient bulbs

Normal incandescent light bulbs are terrible at their job. Sure, they're a big step up from the candles and kerosene lanterns of our ancestors, but if their task is to efficiently and effectively provide light, they're a failure. These devices are great heaters but poor sources of light. Not only that, but with a lifespan of about 1,000 hours of use, they're a short-term item, a throwaway. This translates to less than a year's operating time in a typical living room. They require frequent replacement and almost always end up in landfills. But what are the alternatives?

The primary alternative today is fluorescent lighting, both straight-tube and compact fluorescent. These options use roughly one-quarter of the energy for the same amount of light as incandescent bulbs. I repeat: Fluorescent lighting gives the same amount of light for roughly one quarter of the energy! That's efficiency: getting what you want (light) while using much less energy.

If you still have incandescents in your living areas, my advice is to go out today and start working on replacing them. Start with the most-used lights in your home and find suitable replacements. With compact fluorescents, you can generally cut the wattage by a factor of three, and you'll end up with a bit more light than you started with as soon as the bulbs warm up. Do note that compact fluorescent light bulbs (CFLs) don't come up to full brightness immediately. Some people find this irritating. I personally like the gentle adjustment.

Also be aware that CFLs come in a number of *color temperatures.* These are specified in kelvins (K), and they range from 2,700 K, which is much like incandescent light, to 6,000 K or more, a range that approximates sunlight, although you typically see it as very white, even with a tinge of blue. The whiter bulbs seem brighter to human eyes, though the light measurement is actually similar to more yellow bulbs. Different people have different color temperature preferences — make sure you get what you want.

Compact fluorescent bulbs do contain a tiny amount of mercury, so recycle them at the end of their lives. But bear in mind that if you were using incandescent light bulbs instead, in most cases you'd be putting more mercury into the air by using utility electricity and its not-so-clean sources. And the amount of mercury in a CFL is not alarming. A typical watch battery has six times as much mercury as a CFL.

Another lighting technology is on the rise: light-emitting diodes (LEDs). LEDs are electronic devices that give off light when energized. Quite efficient and extremely durable, these lights are still fairly expensive, however. The light is also directional, so many of these lights aren't yet ready for prime time home lighting. They're great for accent lighting or outdoor lighting, though. A few companies have brought out products that try to approximate the feel (light quality, color temperature, and directionality) of incandescent bulbs in an LED product, but the prices are still high. I encourage you to explore the options and use LEDs where appropriate. Compared to CFLs, the efficiency is higher, and the life span may be ten times as long, which makes these bulbs very convenient.

Installing and switching lights efficiently

How lighting is installed and switched also affects its energy efficiency. Try the following tips:

- **Use fewer shades and light covers.** Hiding lights in cans or behind valances or in thick covers of any kind decreases how much light makes it out to your room. Although bare bulbs aren't terribly aesthetic, the closer you get to a bare bulb where you need the light, the more you'll get for your lighting watts.

- **Put your lights on individual switches.** Putting four lights on one switch means that your choices may be 0 watts and no light or 100 watts and more light than you really need at once. With lights on individual switches, you have control of how much light you actually use and therefore how much energy you use.

- **Use timers, motion sensors, and daylight sensors to make your lighting more efficient.** Sensors and timers can give you what you want without waste. In my home, we use timers for lights outside our shop, where leaving the lights on without noticing is all too easy. With our timers, the maximum waste will be 15 minutes, and only very rarely do I have to go back inside to turn the timer on when I'm working in these areas.

Electronics

Many electronic devices present *phantom loads* — energy use when the device is off — because of poor design. (I discuss phantom loads in more detail later in this chapter.) Most home electronics fall into one of these two categories:

- **Audio/video:** The run-time load is not usually a big deal, but these are frequently phantom loads, so setting up your entertainment centers for switching is a wise move.

- **Computers:** Laptops almost always use less energy than desktops, and measuring the charging wattage with battery full and not full can help you develop a sensible use strategy. If you put a wattmeter on your computer, you can find out how much energy it uses in sleep mode, how much it uses in hibernation, and how much it uses when off; then you can make wise decisions about your computer's power settings.

Computer peripherals, such as printers, scanners, and external hard drives, often present phantoms, so using a power strip with switched outlets is a good control strategy.

Shop for electronics with a wattmeter and discriminate against poor design. If people continue to press electrical store employees for better gear, the word will gradually filter back to the engineers who are making poor design decisions.

Other appliances

At a minimum, buy appliances that have an Energy Star rating. I suggest you go further than that: Compare specific energy use for all appliances and buy the best you can afford within your needs. Be *very* aware of phantom loads (which I discuss later in this chapter) with appliance purchases, and if you can't avoid them, wire in switched outlets to control them.

Here are my specific recommendations for appliances:

- ✔ **Clothes washer:** Front-loading washing machines are almost always more energy efficient that top loaders, using much less water and cleaning clothes better in most cases. Watch out for phantoms here, too.

  There aren't huge differences between dryers — they're big energy users because they're heaters. My first advice is to buy a front-loading washing machine with a high spin cycle; this gets a lot of the water out of your clothes upfront. Then use a clothesline or indoor clothes rack if possible. You'll save energy, and you won't have to clean out the lint trap.

- ✔ **Dishwasher:** Consider the drawer styles that allow you to use one drawer or both so you can wash smaller loads. Dishwashers are frequently phantom loads, so build a switch into your plans.

- ✔ **Microwave:** These are actually quite energy efficient, setting aside the health concerns that some people have. They cook food quickly with a modest amount of energy. Choose one that has minimum programming after being plugged in so you can turn the phantom off between uses.

Cooking is a good example of when you can save energy by using appropriate appliances for appropriate jobs. If you need to cook the Thanksgiving turkey, you need a full-sized oven. But if you just need to warm up last night's leftover Chinese, the microwave will serve well. If you only want a grilled cheese sandwich, use your toaster oven. Similarly, think about not heating up a whole tea-kettle of water if you just need a cup.

Both the American Council for an Energy-Efficient Economy (www.aceee.org) and Energy Star (www.energystar.gov) are excellent resources for appliance efficiency information. Be a careful shopper, and you'll end up with appliances that do their jobs frugally instead of giving you waste along with their services.

Clearing Out the Phantom Loads

A *phantom load* is an electrical load that still uses energy when it's switched off. It's not uncommon for me to find 3 to 5 kWh per day of phantom loads in a typical American home. This is a fertile field for energy savings!

Phantom loads can vary from a tiny 1-watt load on a phone charger to the monstrous 72-watt phantom I found on a stereo tuner last year. The former is only 24 watt-hours per day, which isn't significant in a home that uses 25,000 watt-hours (25 kWh) per day. But the little 1-watt phantoms can add up if you have lots of them, especially if you're off-grid or trying to make all your energy with renewables. The 72-watt phantom is outrageous. That's more than 1.7 kWh per day — for nothing!

These wasteful phantoms serve no purpose; it's just poor design. Your mission, if you choose to accept it, is to find the phantoms in your life and eliminate them.

Finding phantoms

Your best friend when seeking phantoms is a watt/watt-hour meter (check one out in Chapter 6). Common brands are Kill A Watt (www.p3international.com), Watts Up? (www.wattsupmeters.com), and Brand (www.brand electronics.com). These all do the same basic jobs, and each has various other fancy features. For phantom hunting, you just need the watt feature.

If you want to hunt down and eliminate the phantoms, you need to check every plugged-in appliance in your home. Here's what you do:

1. **Plug the meter into the wall.**

2. **Plug the appliance into the meter.**

   Make sure the appliance is turned off.

3. **Scroll to or push the button for the watt function.**

   Some meters need ten seconds to come up with a reading.

If the phantom load is less than 3 watts in a modern on-grid home, I don't usually lose too much sleep over it, unless I see ten or twenty 3-watt loads. If the load is higher, do something about it. I commonly see phantoms of 10 to 20 watts on TVs, VCRs, DVD players, and computer peripherals, just for a few examples.

Extinguishing phantoms

The best place to kill phantoms is at the appliance store — don't buy stuff that presents phantom loads. I literally walk into stores with my wattmeter and plug stuff into it. Then I know before I buy how much energy waste the device will incur.

 Okay, so sometimes you can't find a phantom-free product (I hope that changes) for a specific application. And if you already have lots of phantom loads in your home, you may need some mitigation strategies. Here are my recommendations, from the crudest to the most elegant:

✔ **Unplug it when not in use.** Not very convenient but effective.

✔ **Use a power strip.** This is more convenient and even better if you mount the strip in a place that's easy to reach.

✔ **Add on an external switched plug.** These devices are available in most hardware stores, and they allow you to switch off one device. Just plug the block-with-switch into the wall and plug your appliance into it. I use several of these with a power strip for my computer and peripherals. Then I can turn off my printer, external hard drive, auxiliary monitor, and so on but still charge my laptop.

✔ **Replace your outlet with a switched outlet.** These single-gang devices have one plug receptacle and one switch in a single unit, so they fit in a standard single electrical box.

✔ **Prewire or rewire your home with specific outlets that are controlled by remote wall switches.** I recommend these for your entertainment center, computer center, microwave, dishwasher, and clothes washer at a minimum.

Adding Up Your Efficiency Results

Figure 7-2 shows a typical load distribution (all energy, not just electricity) in North American homes. Although it may not be typical of your home, here are some lessons you can pick up from the largest areas in the pie chart:

✔ Focus on heating/cooling efficiency.

✔ Improve your hot water supply system.

✔ Cut into your lighting load.

✔ Trim all appliances.

After you put the techniques in this chapter to use, you're ready to calculate and celebrate. Review your *before* measurements — perhaps you've set up a spreadsheet. Then take measurements after your energy efficiency measures. Calculate (or let the spreadsheet do the number crunching) to see what you've gained.

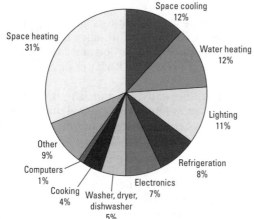

Figure 7-2: Breaking down household energy use.

Space cooling 12%

Space heating 31%

Water heating 12%

Lighting 11%

Other 9%

Computers 1%

Cooking 4%

Washer, dryer, dishwasher 5%

Electronics 7%

Refrigeration 8%

What could an efficiency calculation look like for Joe and Jane Homeowner, whose pre-efficiency energy budget was 105 kWh per day, with electric heat?

✔ **Heating/cooling upgrade:** Joe and Jane got rid of an older electric boiler and replaced it with a high-efficiency ground-source heat pump. This can easily save them 15 kWh per day, averaged year-round. (Results here vary a lot, depending on your current equipment, the size of your house, your climate, and the current thermal efficiency of your building.)

✔ **Insulation replacement:** Joe and Jane's house was poorly insulated. They brought it up to the highest standards to see another estimated 15 kWh per day in savings.

✔ **Building envelope:** Joe and Jane re-sided their house, used house wrap, and had their contractors do a very careful job of sealing their building envelope. Again, quantifying the benefit exactly is tough, but compared to their terribly leaky pre-retrofit house, this can save them 20 kWh.

✔ **New refrigerator:** A new energy-efficient fridge dropped their bill another 2 kWh per day.

- ✔ **Water heater adjustment:** Joe and Jane's water heater was brand new and well insulated, but the thermostat was a bit high at 130°F. They turned it down to 125°F and shaved 1 kWh per day off their load.

- ✔ **Lighting retrofit:** The Homeowners' lighting retrofit was one of the easiest strategies. It lifted 8 kWh per day from their bill.

- ✔ **Appliance replacements:** Upgrading their other appliances shrunk their load by another 4.5 kWh per day.

- ✔ **Getting rid of phantom loads:** 2.5 kWh per day were extinguished with the methods for ditching phantom loads I mention earlier in this chapter.

Some of these numbers have to be guesses; others are easily measurable. But the total is a hard number after you live through a heating season with the new home. The utility meter doesn't lie. In Joe and Jane's case, the home went from using 105 kWh per day to using 37 kWh per day.

Can you get this sort of result? It depends. Joe and Jane started with a truly awful home in terms of energy consumption and pulled out all the stops to improve it. If your house isn't as bad to start with or you can't invest as much in upgrades, your results may not be as dramatic.

The more you can do to reduce your load, the more attainable and satisfying your wind-electric system will be. Mine this deep vein of energy. Many of the strategies I mention in this chapter are very low cost but high return. Others take more investment and need more time to pay off. Implementing them all will make an enormous difference in your personal energy picture.

Chapter 8

Determining Your Site's Wind-Energy Potential

In This Chapter

▶ Studying local effects of the wind

▶ Looking at potential turbine sites

▶ Checking wind speeds with different tools

*Y*ou can hug trees, dive into water, and stick a shovel into dirt. Wind, however, is much more difficult to wrap your arms — and your mind — around. At its simplest, wind is moving air. You can feel it moving past you; you can lick your finger and stick it in the air to determine the wind's direction, and in that way, you can sense even very slight breezes. On the other end of the scale, you may experience winds so strong that they blow you down the sidewalk or move your car sideways on the freeway. But you can't see it, taste it, touch it, or smell it.

Wind is the driver behind wind-electric systems, so understanding what it is, how it works, and how to assess its potential on your property is crucial to good system design. A variety of methods are available to estimate wind speed on your site. Some are very subjective; others less so. None are perfect, unless your budget is unlimited. I talk about many of these methods in this chapter and help you come up with a plan to use them well in your situation.

Understanding Wind's Local Effects as Part of the Global Engine

My colleague Paul Gipe has such a gorgeous paragraph about wind in his fine book *Wind Power: Renewable Energy for Home, Farm, and Business* that I thought I'd borrow it here:

"The atmosphere is a huge, solar-fired engine that transfers heat from one part of the globe to another. Large-scale convective currents set in motion by the sun's rays carry heat from the lower latitudes to northern climes. The rivers of air that pour across the surface of the earth in response to this global circulation are what we call wind, the working fluid in the atmospheric heat engine."

This global engine leads to global wind patterns, with areas of predictable flow and areas of predictable calm. But for wind-electric purposes for your home, these global wind patterns are irrelevant beyond scientific curiosity. The more important information is how wind works close to home.

The air pressure differences you hear about on the nightly news are part of the global wind engine. Air flows from areas of high pressure to areas of low pressure. Local topography affects this flow in a variety of ways. Several specific local effects are important to small wind systems, and I discuss them in this section.

The factors and geographic features that give you wind also provide a certain level of predictability. One aspect is the direction of the wind, which affects where you should site your turbine. In any given location, you find a *prevailing wind* or perhaps two or three predominant wind directions. See the later section "Visualizing wind speed and direction with wind roses" for details.

Understanding patterns due to uneven heating

Certain features, such as large bodies of water and mountain valleys, can cause an area to heat and cool unevenly. This section introduces some daily wind patterns that occur as warm air rises and cool air sinks as the sun makes its path across the sky.

Onshore and offshore winds

The typical wind pattern near the ocean and other large bodies of water is to see an onshore breeze in the morning. Every day, the sun comes up and heats the water and land. Water heats more slowly because it has a higher *specific heat capacity* (it can absorb more heat per volume without changing much in temperature). The land heats up relatively quickly, warming the air above it; that air rises, flowing up the land and away from the cool air over the ocean, creating an onshore breeze.

In the afternoon and evening, the land and water cool down, but the land cools down more quickly; winds tend to fall off the land and blow offshore. Many coastal dwellers see this pattern (shown in Figure 8-1) — onshore in the morning and offshore in the afternoon — repeat itself day after day after day, with higher intensity when the land warms more and when regional or global patterns intensify the local patterns.

Local topography and conditions modify this pattern, so it's not always consistent. For instance, nearby islands affect direction and intensity of wind, and larger-scale prevailing winds do as well. (For more on the effects of topography, see the later section "Looking at the influence of the shape of the land.")

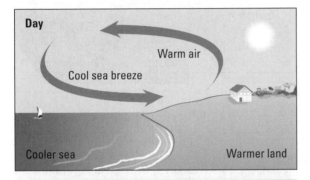
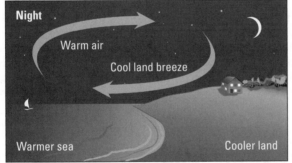

Day

Warm air

Cool sea breeze

Cooler sea

Warmer land

Night

Warm air

Cool land breeze

Warmer sea

Cooler land

Figure 8-1: Onshore and offshore breezes are predictable.

If you know you have a pattern of onshore and offshore breezes, it may affect where you site your turbine. In all cases, you want to get well above all obstructions. And if you know the direction and intensity of most of your wind (see the later section "Visualizing wind speed and direction with wind roses"), you can position your tower to take the best advantage of the wind you have.

Up- and down-valley winds

In mountain valleys, a pattern very similar to onshore and offshore winds occurs: Heating of the land encourages warm air to rise and flow up the valley in the first half of the day, and the air falls back down late in the day and during the evening as the land cools (see Figure 8-2). These effects are most obvious in the summer, when solar heating is at its peak.

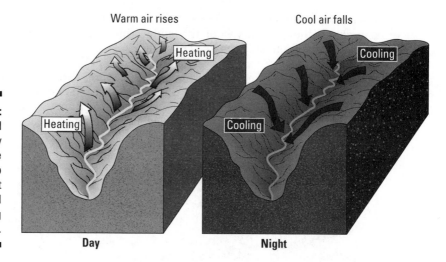

Warm air rises   Cool air falls

Heating   Cooling

Figure 8-2:
Up- and
down-valley
winds are
due to
different
heating and
cooling
patterns.

Heating   Cooling

Day   **Night**

In general, the winds in valleys are more intense in the middle of the valley and weaker on the sides of the canyon. In some places, the prevailing winds coincide with these local effects, such as when the prevailing wind is up-valley.

Understanding how the winds work on your site will help you position your tower. If you have the whole valley to play with, you'll want to put your tower in the middle. And don't assume that just because you live in a valley that you will have this pattern. Some valleys have little wind, especially gradual valleys in the middle of tall mountains.

Looking at the influence of the shape of the land

Every property has its own *microclimate* (the specific climatic conditions on your property, which differ from other places in your neighborhood or area), and this includes wind. Here are a few ways in which the shape of the land and human-made obstructions can influence local wind patterns:

✔ **Tall obstructions such as trees:** I can remember burning stumps when we were creating our 1-acre clearing in the forest. The smoke gave a very graphic image of the air flow within the bowl of the clearing, which was surrounded by 120+ foot tall trees. It gave good lessons about how wind deflects off trees and rolls around with turbulence. I'm starting to sound like a broken record, but the answer to the problem of tall obstructions is (drumroll) . . . a tall tower. If you have only a few trees, you'll want to move your tower well away from them to reduce turbulence and get to stronger winds.

✔ **Ridges:** Another local effect is potential increases in speed over ridges, where you may see 50 percent more wind than on the way up the slope or on the way down the slope on the other side. This isn't something I'd make exact predictions for, but placing a tower on a ridge top is definitely a good strategy.

Smooth ridges are good, but sharp edges and steep slopes tend to create turbulence. Gaps in ridges are classic wind sites — for example, the passes in California.

If you get a chance to tour a wind farm, you may be interested in seeing the placement of the turbines. The spacing and the exact locations of these amazingly huge machines isn't exactly intuitive. Wind farmers have the benefit of detailed wind studies that indentify the spots with the best resources. By understanding the local microclimate, you can make educated guesses toward the same goals.

Identifying Potential Turbine Sites

After you understand how topography affects the wind (see the preceding sections), the next step is to identify a potential tower location, which I discuss in this section. Then you want to get a good estimate of the annual average wind speed at turbine height on that site. (For info on types of wind speed, see Chapter 5.)

Later, in "Analyzing Wind Speed with a Variety of Tools," I discuss how to measure and estimate wind speed. Take your time with this process. You're not looking for an easy answer; you're looking for the most accurate answer you can come up with, leaning toward the conservative side. Remember, the accuracy of this number affects your complete system design and therefore the cost, the performance, and your satisfaction in the end.

Living on-site with your eyes and ears open

Living on-site can help you avoid serious siting mistakes. So if possible, live on a site and experience the normal weather pattern. This can give you subjective but valuable information about the overall wind resource, the peak winds and their frequency, and the direction of winds. Taking some notes on a regular basis will help you remember what you've observed through the seasons. Doing so also gives you the opportunity to implement some of the wind analysis tools in this chapter.

If you're contemplating installing a wind turbine on a site you don't live on, you're taking a risk if you lean on only occasional observation. Perhaps your visits have been during windy times that are actually quite rare. You'll be very disappointed if you install a turbine based on partial observation. If you can't live on-site, visit often, rely on neighbors' observations, and use all other available tools.

Understanding the big picture of your area

Wind has a direct relationship with land. The closer you are to the land, the lower the wind speeds; the farther away from the land you get, the higher the wind speeds, and therefore the higher the wind energy potential. Trees, buildings, and other stuff on the ground slow the wind (see the earlier section "Looking at the influence of the shape of the land" for details). Taking this general information to your specific property can lead you to a better understanding of what you're up against in terms of tower placement, tower height, wire run, and other design parameters. If you own a small lot, you need to look beyond its borders to the neighborhood. If you own a large ranch, focus in on your home site.

A topographic map (such as those available from the U.S. Geological Survey) of your area can come in very handy at this stage. Is your property all flat, all sloped, or complex? What about your neighbors' land? Are you in a valley, on a ridge, on the highest point, beside a lake or ocean, or in the middle of a prairie? Getting a bird's eye view of your neighborhood can help you come to conclusions about your wind-electric system plans.

Some sites are very simple, open, and obvious. Others are difficult to picture, with the complexity of hills, valleys, varied forests, and so on. Take a look from high vantage points, maps, aerial photographs, and any other views you can get. Get comfortable with the big picture so that when you're walking the ground among the trees, you still have a sense of where you are in the topographic scene.

Start high: Climbing something tall and looking

All other things being equal, the highest spot on your property is likely the best wind site. It allows you to get farthest away from the earth with a shorter tower, and you gain the advantage of any wind acceleration over a hill.

This spot isn't always the best spot for your tower, though. It may be far from your home site or from utility connection. Running electrical lines to this site may be difficult, expensive, or unaesthetic. Or perhaps it's close to your neighbors' property, which may involve legal restrictions on placement or aesthetic, visual, and audible impacts you don't want to impose on your neighbors.

But start by checking out this high spot. One wind-analysis strategy is to climb something tall and look around you. As a climber of trees and towers, this is one of my favorite ways to look at a wind site. And it's not just because I like the thrill. Sure, the views are great, and the freedom of being up in the tree tops or beyond is unparalleled. But there's solid, practical usefulness in getting *above* your site to give yourself a clear view of the topography, the vegetation, and the property layout.

I like to get as high as possible in a tree that's central to the property or on a high point. Using a long tape measure, I identify my height above the ground. I take digital photos of eight compass points (north, northeast, east, southeast, and so on). And I use a site level (pea shooter) to look at nearby and distant obstructions to get a baseline for tower height. I look in extra detail toward and away from the prevailing wind(s), because these directions are more important. (See the later section "Visualizing wind speed and direction with wind roses" for info on identifying prevailing wind.)

I normally take digital video of a full 360 degree view, with commentary, so I don't have to remember or take written notes in a tree top. This gives me clear information when I'm back in my office or down on the ground talking to my client.

Take your time and work safely when climbing and working aloft (see Chapter 17 for important guidelines). I want you to live to enjoy your wind energy system. If you're not up for climbing, consider finding a local arborist to go up for you. Then you can experience the scene through the camera lens and have a valuable perspective on your property.

Some properties are situated such that a nearby road, ridge, tower, or even building can give you a better perspective than you have on the ground. Take advantage of these local vantage points, too, and add the information you gather to your stash of perspective.

You can combine information you collect from the high spot with the data I discuss later in "Analyzing Wind Speed with a Variety of Tools." Knowing where the local obstructions are can help you decide how high your tower needs to be and what to expect of your site compared to whatever data you have.

Figure 8-3 shows a map labeled with site structures, vegetation, obstructions, and potential tower sites.

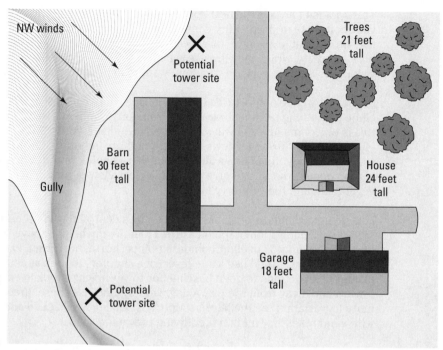

Figure 8-3:
A sample
site map.

Considering nearby topography and local restrictions

After you review your own property, look at the rest of the neighborhood. Does the land drop away from you in all directions? Or are you in a bowl? The *30/500 rule* (which says that a wind generator's lowest blade should be at least 30 feet above anything within 500 feet; see Chapter 14) is really a minimum standard for level terrain. If you're in a bowl, you need to go much higher to get up into the clear wind.

Where's the prevailing wind in relation to neighboring obstructions? Your top priority is a clear view to the prevailing wind, but having clear exposure downwind is also important. Obstructions behind a wind turbine back up the wind and slow it down, causing decreased wind speeds (read "decreased energy") and increased turbulence. A wind rose, a type of graphic I discuss later in this chapter, can help you figure out where most of the wind comes from most of the time.

Reconsider your proposed sites in view of the neighboring terrain and potential tower heights in each location. Measure distances from the proposed sites to your property lines, to your home, and to the utility connection(s).

And ask yourself whether installing your wind-electric system on a particular site would be desirable, practical, and affordable.

Legal, social, and safety issues come into play when you're siting a turbine. You may need to deal with some or all of these as you plan your system. Knowing what you're up against in advance can save you money, time, and heartache. Check with your local jurisdictions for restrictions on tower height and *setbacks* — the distance that your tower has to be from buildings and property lines. These laws are often great impediments to good wind-electric system design. I explain how to find zoning rules in your area in Chapter 2.

If the highest point on your property isn't suitable for your turbine, look at other potential sites on the property. Weigh the pros and cons of each site and make a prioritized list of the possibilities. You also need to consider which type of tower you want to use (see Chapter 14) and what its space and other requirements will be.

Analyzing Wind Speed with a Variety of Tools

Your goal is to estimate an average wind speed for your site, preferably estimated at *hub height* — at the height above ground where you hope to install your turbine. (For info on choosing a turbine height, see Chapter 16.)

You can't just guess about average wind speed — that's a bad idea. Guessing here is about as wise as guessing your energy usage. Your ultimate turbine sizing choice depends on your energy need (kilowatt-hours per day; check out Chapter 6 to determine this need) and your wind resource (average miles per hour). If you guess at both of these numbers, the turbine sizing will be a complete guess as well.

So you need to find some methods to do better than a guess. There's no simple, cheap, and accurate method to determine average wind speed, so I have a number of tools for you to add to your toolbox. I hope that the combination of several of the following tools can give you a decent estimate, or at a bare minimum, an educated guess about your average wind speed. Analyzing wind speed may take several hours on a simple site, or it may involve research and/or measurement over several days or even months.

Finding an experienced wind site assessor is perhaps your best option. These folks will use these tools and perhaps more to give you an estimate of your resource. Educating yourself is harder and more time consuming, and you'll have to ask yourself whether you trust your assessor in that situation.

Taking direct measurements

In theory, taking direct measurements should give you the most accurate, objective data to help you in your wind estimates. However, costs are usually high, and you have to measure for a long time to make sure you get a good average. However, if you're willing to spend the money, setting up an anemometer or a small test turbine can give you some useful data.

Using an anemometer

An *anemometer,* shown in Figure 8-4, is a common wind measuring device. The most common type has three cups that look about like half ping-pong balls. They spin around on a vertical shaft, and the shaft is connected to a logging device of some sort that may record instantaneous, peak, and average wind speed (see Chapter 5 for more on types of wind speed).

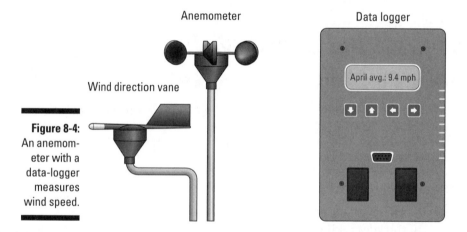

Anemometer

Data logger

April avg.: 9.4 mph

Wind direction vane

Figure 8-4:
An anemometer with a
data-logger
measures
wind speed.

In years past, people used to say that you needed to put up an anemometer at hub height and record the wind resource for a year or more. This is part of what utility-scale wind prospectors do, so it's not surprising that people tried to apply this methodology to home-scale wind. And it wouldn't be an entirely bad idea if it were inexpensive and quick to do it, but it isn't. Heed the following warnings:

✔ **Cost:** I asked one wind energy expert about the cost to do a proper job of resource analysis with anemometry, and he said it'd cost between $10,000 and $20,000. That includes installation labor and data analysis, so you may be able to do it for less. But even if you can do it for $5,000, it's a stretch to recommend that for a wind-electric system that may cost in the $40,000 to $60,000 range.

You may be able to find a simple, cheap, almost homebrew anemometer (an example is a bike speedometer calibrated to an anemometer head) and put it on an inexpensive mast. If you can get it up above obstructions and get good data cheaply, go for it!

✔ **Time:** Do you want to wait a year before doing something about your wind-electric system? And is a year enough? If that's all the data you have, it's a bit shaky to stake your design on. What if the year you measure is well below or above average? You may make a poor design decision based on a freak year. Monitoring for several years would really be ideal if an anemometer is the only tool you're going to use. Fortunately, it isn't, as you find out later in this chapter.

If you have the opportunity to do *some* on-site monitoring with an anemometer at a reasonable cost, it may be very useful, in combination with other tools, especially with other local data that I describe later in this chapter.

The tool isn't an "anenometer"; it's an *anemometer.* Remember the pronunciation by remembering that your "mom" is in the middle. The word comes from the Greek word *anemos,* which means "breath" or "wind."

Setting up a mini test turbine with a watt-hour meter

As soon as people realize how expensive it is to put up an anemometer, one common idea is to install a small wind generator instead. For not that much more money, you can install a *sailboat turbine,* one of the micro turbines with diameters of perhaps 2 to 5 feet. This turbine won't make you a lot of energy, but combined with a watt-hour meter (check one out in Chapter 6), it can give you an idea of what your wind energy potential is.

The idea of setting up a mini test turbine has several challenges:

✔ You still need a tall tower to get well above the ground and other obstructions.

✔ The tower for a small machine won't support a larger machine. Each tower can handle only so much thrust, which translates into a certain amount of swept area (the area covered by the spinning blades). If your test yields good results, you'll need to start over with a tower for a home-scale machine.

✔ Your metering will give you good numbers, but they'll be watt-hours — energy produced. To understand what that means in terms of average wind speed (which you need so you can estimate production from larger machines) takes some extrapolation and will probably mean you have to trust the production figures from the small-machine manufacturer.

All that said, this can be a useful method, especially if you're of a mind to play with wind energy but aren't sure how deep you want to go. Mini turbines can give you info on how the wind works on your property; what a wind turbine looks, sounds, and feels like; reliability; neighbor's reactions; and so on. If you go into it with an I'm-going-to-discover-something attitude, you'll discover something, and you'll probably have a good time doing it.

Collecting other people's data

Some information on wind resources is yours for the taking — provided you can track it down. In this section, I discuss visual representations of wind resources, such as wind maps and wind roses, and I talk a bit about collecting weather numbers from various resources.

Consulting wind maps

If you'd asked me ten years ago, I would've said that wind maps were a questionable tool for determining the wind resource on your property. At that time, for most places, the data wasn't particularly good. Today, the picture is different. Although I'd still be cautious about using wind maps as your primary or sole source of information, the wind mapping available today is of much higher quality, and with some interpretation, it can be a useful tool if you understand the limitations.

Several wind map sources are available; the main one I use is AWS Truewind. You can find their ***windNavigator***® at `navigator.awstruewind.com` (see Figure 8-5). The free portion of their service allows you to input a street address or longitude and latitude. The display then shows you a topographic map with the location and provides generalized wind data.

The data that ***windNavigator***® presents is the average for a 2.5 km grid, and you have the option of looking at data from heights of 60, 80, and 100 meters. This Web site is obviously aimed at utility scale wind resource assessment, because even the lowest height — 60 meters — is above where most home-scale wind generators will live. But it's still a point of reference, and you can correlate it.

Suppose your address has an average wind speed of 4.13 m/s (9.25 mph) at 60 meters. Using a *wind shear calculator* (which adjusts the wind speed prediction based on surface roughness), such as the one at `www.windpower.org/en/tour/wres/calculat.htm`, you can calculate the projected average wind speeds at heights lower than the 60 meters given by the wind map.

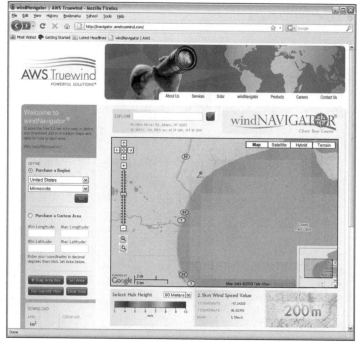

Figure 8-5:
AWS
Truewind's
*wind
Navigator®*.

Courtesy of AWS Truewind, LLC

I recommend that you test the wind map in your area by doing this same pro-cess for a site with a known average wind speed, such as a local wind-electric system or a wind monitoring station. Then you can get an idea of the accuracy of the wind map in your area and be able to extrapolate with more accuracy. Round *down* in this process — you don't get any advantage from exaggerating your wind resource.

Visualizing wind speed and direction with wind roses

Wind direction is described by where the wind is coming from. In my location, for instance, the clear prevailing wind is out of the southeast. But we also have two secondary prevailing winds, from the southwest and the northwest.

A *wind rose* is a graphic illustration of the frequency and energy potential in the winds from each direction (see Figure 8-6). These are available from some wind mapping services or databases, and they can be very handy when look-ing at your property and its wind resource.

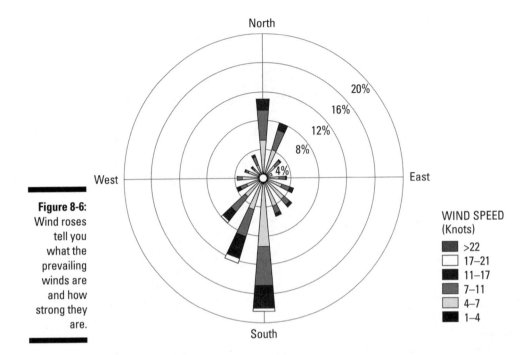

North

20%

16%

12%

8%

4%

West

East

WIND SPEED
(Knots)

■ >22
□ 17–21
■ 11–17
■ 7–11
□ 4–7
■ 1–4

South

Figure 8-6:
Wind roses
tell you
what the
prevailing
winds are
and how
strong they
are.

A wind rose typically has three basic components:

✔ The basic framework is a compass diagram, with the directions labeled.

✔ Superimposed on this are one or two data sets. The primary information is the frequency of wind blowing from a certain direction (note that it is *from*, not *to* the compass point).

✔ Some wind roses also show wind speed classes, indicating the percentage of the time the wind is averaging a certain range of wind speeds.

Wind roses can be a handy, graphic way to quickly reference the prevailing winds. Adding this information to your analysis can help you decide whether obstructions such as trees or buildings are a deal-breaker for your wind-electric system design. In other words, wind roses can show you prevailing winds so you know where obstructions have the most impact.

Sifting through local weather data

Local weather data is a source of information that you should seek out diligently. Wind is a local phenomenon. Knowing what the wind is like 100 miles away is fairly useless. Knowing what it's like 10 miles away may be useful. Knowing what it is a mile from your site may be the best information you get.

Common sources for local weather data include the following:

- Government sources, including the National Weather Service (NWS), National Oceanographic and Atmospheric Administration (NOAA), National Renewable Energy Laboratory (NREL), and others

- Local and regional airports

- Local and regional news organizations, including newspapers, radio, and television stations

- Local weather hounds — the folks in your community or region who have a hobby interest in weather and keep records from their own anemometers, thermometers, rain gauges, and such

- Wind-electric system owners with anemometry

It's unlikely that any of this data is perfect unless your next-door neighbor has had an anemometer for 10 years and is very particular about keeping the data. But any of this data can be useful as you try to figure out what your wind resource actually is.

I recommend gathering all the data you can and identifying the specific source as closely as possible. Some data is more valuable than others. For example, airport data is often taken very close to the ground, and airports are generally sited to avoid wind. I've seen airport anemometers that are 10 feet off the ground and shielded by trees and obstructions for more than half of the wind directions.

Knowing at what height your local wind data was taken and what the surrounding obstructions are is crucial to interpreting it. My advice is to take the best data you have, compare the site characteristics where the data was taken to your site's characteristics, and then make an educated guessti-mate of your average wind speed. But round *down* in general, not up. You'll be much happier if your surprise is that you make more energy than you expected, rather than less.

A log for a site may look something like Table 8-1. You may also want to collect data on peak wind speed so you have an idea of how durable your turbine needs to be.

Table 8-1		A Sample Log of Local Wind Data			
Source	*Distance from Site (miles)*	*Average Wind Speed (mph)*	*Measurement Height (meters)*	*Highest Obstruction within 500 feet (meters)*	*Notes*
NOAA	17 miles	9.4	30 m	24 m	Consistent site
Airport	6.5 miles	7.8	10 m	30 m	Shielded to secondary prevailing wind
Local paper	4 miles	8.4	10 m	10 m	Mostly open field
Local hobbyist	1.3 miles	9.6	25 m	18 m	Wooded but consistent

In this case, the NOAA data is from the tallest tower, but it's farthest away. The airport data is very questionable because of both height and obstructions. The local paper data is at a low height, but it serves to confirm the general resource in the area, especially when compared to the local hobbyist's data. The hobbyist's data is the most useful because it's closest to you, and it's also on a reasonably tall tower with some clearance above the trees.

Looking at a topographic map, you can compare the two most local sites to yours and gauge whether these sites are similar in exposure to your own site. If the whole area is level and consistently vegetated, the local numbers may be very good indications of the wind resource on your site. If you're on a high point relative to the other sites, you may expect a somewhat better resource.

Looking and listening

You can get some fantastic information about your wind resources simply through personal observation — your own or that of your neighbors. In this section, I discuss some subjective tools for evaluating the wind. Use them in conjunction with some of the harder data I introduce in the preceding sections.

Connecting vegetation to average wind speed and direction

At first glance, vegetation data seems like a throwaway. But don't toss nature out so quickly. You're trying to quantify the wind resource. Maybe the wind itself has left you clues.

Wind blows in regular patterns, from specific directions in each season or in each type of storm. The effects of the wind add up and show themselves in the vegetation in the area. If it always blows hard from one direction, the trees become deformed, or _flagged,_ in the opposite direction. This deformation can be very severe in high-altitude environments. I've seen carpeted vegetation in New Hampshire's White Mountains, where the wind doesn't allow the vegetation to grow higher than a few feet; that is, the wind blows the vegetation down to the ground so that it grows horizontally, not vertically.

This effect on vegetation has even been quantified sort of scientifically. The Griggs-Putnam Index (see Figure 8-7) identifies various levels of flagging and correlates them to average wind speed.

Some caution is in order when using this method. Different species of trees grow and exhibit flagging differently. The Griggs-Putnam Index focuses on flagging of conifers — evergreen trees with year-round needles. But even within this general category, there are variations. For instance, in the Pacific Northwest, you have Douglas-Fir trees, which frequently grow an individual branch or two pointing out in an odd direction, oblivious to the prevailing winds.

Looking at vegetation patterns for wind clues is subjective, and applying it takes experience. Sometimes flagging is very pronounced and obvious, and other times it's more subtle. Don't look at only one specimen; look at a range of trees on your property. Look from various angles. Check out the flagging in the area when you're aloft in trees or other high points. Also be aware that flagging of older trees may have taken place years earlier, and would not take place again now that recent shelter has grown in the area. Don't use vegetation as your only measure, and compare and contrast the results you get with other information you've gathered about the local wind resource.

Talking with neighbors

Is talking to neighbors a subjective tool? You bet! Is it useful? It can be. I like talking to more than one neighbor, and I like to spend enough time to get to know their style. Some people naturally exaggerate, others tend to minimize everything, and some find a balance in the middle. If I can find someone in the middle or average the extremes, I can gather some useful information.

Do the trash cans blow down the street? Did you plant a windbreak back in the 1950s to protect your garden? Do hay bales blow off the top of the stacks up on the hill? These sorts of questions and more can give you some idea of the site. I also ask neighbors whether they know who may be harboring local weather data and get their impressions of the prevailing wind directions.

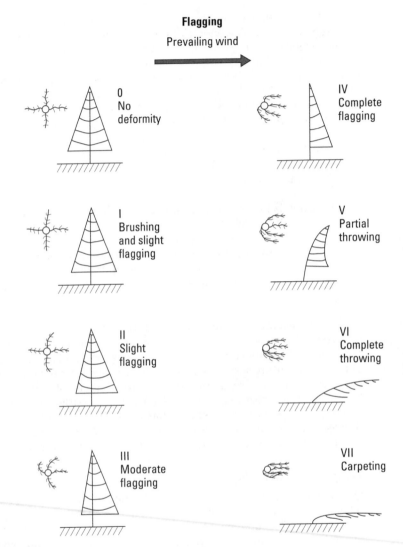

Flagging

Prevailing wind

0
No deformity

IV
Complete flagging

I
Brushing and slight flagging

V
Partial throwing

II
Slight flagging

VI
Complete throwing

III
Moderate flagging

VII
Carpeting

Figure 8-7: The Griggs-Putnam Index links tree flagging to average wind speeds.

Average wind speed	I	II	III	IV	V	VI	VII
mph	7-9	9-11	11-13	13-16	15-18	16-21	22+
m/s	3-4	4-5	5-6	6-7	7-8	8-9	10+
Km/h	11-14	14-18	18-21	21-25	25-29	29-32	36+

The very best neighbors to talk with are those who already have wind genera-tors, though these may be rare. If you're new to this game, you may think that nobody in your area is using wind energy. Perhaps that's true, but it's more likely that one or more wind generators are hiding out on a remote property within range of yours. Get connected to the local renewable energy network by talking to local dealers and enthusiasts and find these gold mines of infor-mation. They can save you a lot of legwork, and you can find out about more than just the wind. These folks have feedback on equipment, installation, con-tractors, and the local bureaucracy.

Chapter 9

Defining Your Relationship to the Grid

. . . .

In This Chapter

▶ Checking out batteryless grid-tied systems

▶ Considering on-grid systems with backup

▶ Living off the grid

▶ Making your decision

. . . .

*W*hen you decide to use wind electricity, you have a number of decisions to make. One important decision is how to be related — or not related — to the utility grid. Here are your basic options:

- ✔ **Batteryless grid-tied:** This system is the most efficient and cost-effective option, but it gives you no outage protection.

- ✔ **On-grid with backup:** This gives you the best and worst of both worlds: outage protection and the ability to get credit for your extra energy — plus the hassles that batteries entail.

- ✔ **Off-grid:** With this option, you rely completely on your own systems, which means you have batteries and probably a backup generator.

In this chapter, I talk about each system's pros and cons, and I raise lots of questions about these options. Your answers to these questions can help you choose a wise system design that serves your needs. Flip to Chapter 3 for an introduction to all the components of these systems and to Chapter 16 for the full scoop on sizing.

Simple and Efficient: Batteryless Grid-tied Systems

Batteryless grid-tied systems are the simplest, most efficient, and most environmentally friendly wind-electric systems. They have only one real drawback — no backup — and many advantages. In this section, I list the advantages and disadvantages of these systems, and I describe the configuration's components and sizing.

Batteryless grid-tied systems have the least owner input and are the least hassle. If you can accept the one major drawback (no backup), these systems are all benefit — do it if you can!

The pros: Simplicity, cleanliness, and cost effectiveness

Batteryless grid-tied wind-electric systems have one job and one job only: to put as much wind electricity onto the grid as possible after servicing whatever loads you have running.

Advantages of a batteryless grid-tied system include the following:

- **High efficiency:** There's no battery bank, so you don't waste energy in keeping the battery charged up waiting for an outage. Nor do you have the cost and trouble associated with purchasing, maintaining, or replacing the battery bank. So these systems are very efficient with your wind energy — the maximum amount gets used by your house and sent back to the grid, creating a credit for you against times when it's not so windy.

  Because batteries aren't involved, these systems tend to be high voltage, which also reduces the wire sizing and increases the efficiency. Systems that run upwards of 200 volts are common. And a batteryless grid-tied inverter will maximize the wind turbine's output by *maximum power point tracking* — loading the turbine optimally to get the most out of it at each power level. This capability is relatively rare in battery-based systems today.

- **Full use of all the energy produced:** The system doesn't throw away any energy on batteries (as in battery-based grid ties) or when batteries are full (in off-grid systems).

- **Flexibility in system sizing:** After you've ruled out batteries, you can size your system by the total energy you desire, total financial return (through energy offset and incentives), or total budget available. (I discuss sizing in more detail later in this chapter.)

✔ **Fewer components:** Without batteries, you don't need a controller (that function is built into batteryless inverters), and in rare cases, you may not need a dump load, either, when the turbine can run unloaded in grid outage conditions.

✔ **Fewer hazardous material and pollution concerns:** These systems bypass the environmental concerns with batteries, which have lead and acid that you need to guard from spillage or exposure to people then recycle at the end of their lives. Batteryless grid-tied systems usually don't have backup generators either (if they do, they must be independent of the wind-electric system), so they also avoid the troubles associated with fossil-fueled generators.

All in all, tying into the utility grid is the way to go unless you have a strong need or desire to be off-grid or have battery backup. And it's the most common system installed in North America today, so you'll be in good company if you choose this, because this is the growth segment of the home-scale wind industry.

For all its faults, the utility grid is a surprisingly good way to move energy about, with no oil spills or nuclear disasters attributed to the distribution network. Why not green it up with renewable energy? Adding wind generators and solar-electric arrays and such all over the grid — from downtown to the rural fringes — will only improve the grid's reliability while reducing the need for upgrading transmission infrastructure. Connecting your wind-electric system to the grid makes your system more efficient, more reliable, and more cost effective. It also means that as far as your loads and usage go, you'll be "just like downtown" — with no need to change your lifestyle or habits.

The cons: When the grid is down, you're out

You may think that batteryless grid-tied systems are too good to be true, but keep in mind that they have a couple of disadvantages:

✔ **No backup:** When the grid's down, you're in the dark. These systems must have the grid to function at all — no matter how windy it is. Without the grid, these systems can't operate — they'll shut down, and they won't deliver any electricity to your loads. If you want backup for utility failures, you need batteries.

Are you dreaming of having a batteryless system but using it when the grid is down? Think again. The wind is variable, and unless you want your computer to drop every time the wind calms down, you don't want to go there. In addition, the inverters just aren't designed to produce any output without a utility signal present. It's part of the UL standard that they have to live up to — it protects line workers, and it protects your appliances. Again, if you want utility outage protection, you need batteries or a generator.

✔ **More bureaucracy and red tape:** With all grid-tied systems, the state electrical agencies as well as the utilities get into the act with their hands out and the orders streaming from their mouths. This extracts time and money from your life — though after you've jumped through the hoops, the bureaucracy usually doesn't have much of an ongoing impact.

If others haven't plowed this ground already, you may need to educate local inspectors, utility engineers, and government bureaucrats. You'll have your work cut out for you (but those who follow you will thank you if you do a good job educating the bureaucrats, who typically don't know much about these systems).

The basic components and configuration sizing

The basic components of a batteryless grid-tied system include the following:

✔ Wind turbine

✔ Tower and foundation

✔ Transmission wiring and conduit

✔ Inverter

✔ Power conditioning equipment and dump load

✔ Utility interconnection equipment

Without batteries, you eliminate one major sizing issue, and life is simpler. All production heads for your energy loads and the grid, spinning your meter in the right direction as often as possible.

All on-grid systems allow you to design without focusing on energy loads, but loads are still a good place to start. Knowing what the total daily load is in kilowatt-hours gives you a baseline. If your goal is to be *zero energy* — which means your system makes as much over the year as your home uses — then the total daily load is your target for system production.

However, being zero energy can be tricky. In many cases, if you produce more than you use, the utility will pay you little or nothing for it. And if the reckoning period is monthly instead of annually, this can be a bad deal indeed. Find out in advance the details of your net metering agreement so you can strategize how best to design the system. In some cases, it may be wiser to design for 60 to 90 percent of your energy load and not risk giving away a surplus. On the other hand, some people are happy to green up the grid, just as they're happy to plant a tree with no financial return.

But when you're on-grid, being zero energy isn't necessary, and your design goal can be one of the following:

✔ To produce some fraction — from 10 percent to 90 percent — of your home's electricity

✔ To zero your bill, which, including all incentives (such as subsidies and tax breaks), may mean that you have a much smaller system than a zero-energy system

✔ To fit within a certain budget

On-grid systems have all these possible targets and the comfort of having any shortfall covered by the grid. After you figure out your own target, you can size your wind generator accordingly, based on your average wind speed and predictions of various wind generator's production. (Chapter 13 has information on wind generators.)

Sizing with batteryless grid-tied systems involves two issues, which are usually combined. You need to decide how big the wind generator is, which comes back to how much energy you want to generate. And you may need to size the inverter, though this is usually sold in a package with the wind generator. See Chapter 16 for more on sizing.

In Both Worlds: Grid-tied Systems with Battery Backup

A grid-tied system with battery backup is similar to an off-grid system (which I discuss later in this chapter) in design and configuration. It uses batteries, so you have some independence and reliability, but it also has the advantages of plugging into the grid. In this section, I cover the pros and cons of this system, along with its components and sizing.

The pros: Reliability and ability to sell

The advantages of grid-tied systems with battery backup are these:

✔ **The grid can absorb all your surplus energy, creating a credit with the utility on windy days that you can then draw from on calm days.** The grid then acts as a 100-percent-efficient (to you, not to the grid) "battery" to store your surplus. State and utility policies vary, but most states mandate that utilities must accept your renewable electricity. This ability means your precious wind electricity doesn't go to waste when your battery bank is full and you're making more than you're using.

✔ **You aren't required to make all your own electricity, so you have flexibility in sizing your system.** If you can't afford a system big enough to make you a zero-energy home, you can make one-half, one-third, or only one-tenth of your energy from your wind-electric system. The grid is there to pick up the balance.

✔ **You have as much electricity as you want, almost all of the time.** While you may need to conserve during grid failures, your electricity isn't subject to periods of calm, dark, or equipment failure while the grid is up. People who've been born, raised, and who now live on the grid come to see this as normal. (If you live off-grid for a while, you may begin to appreciate this a bit more.)

✔ **These systems give backup for critical loads.** You don't need to over-size your system's generating sources or battery bank for times of low-energy weather. Nor do you have to have a backup generator, unless your backup needs are large or critical. This saves money and headache.

The cons: Inefficiency, modest backup, and cost

Having a battery-based system on the grid isn't all a bed of roses. This choice comes with some of the disadvantages of both off-grid and batteryless grid-tied systems:

✔ **Inefficiency:** Batteries introduce inefficiency into the system, using 5 to 40 percent more wind electricity than batteryless systems.

✔ **Only modest backup:** A battery is a limited storage device, and it's an expensive one in more ways than one (as I explain in the later section "The cons: Costs, maintenance, and wastefulness"). With a battery bank that you can afford, you'll be lucky to back up *all* your modern home's loads for more than an hour or two.

More typically, these designs back up only truly critical loads. These may include a few lights in key rooms; phones, radios, and other communications equipment useful during emergencies; perhaps your refrigerator (though this is a significant load); and other key small appliances. How much backup you get depends on watts times hours — watt-hours.

✔ **Battery cost and upkeep:** Battery-based grid-tied systems have all the battery purchase, maintenance, repair, and replacement costs of off-grid systems. The battery bank may be smaller, but it still needs attention and care from you and your pocketbook.

✔ **Bureaucracy:** As with the batteryless grid-tied system, you encounter more bureaucracy and red tape.

The basic components and configuration sizing

Battery-based on-grid wind-electric systems are almost as complicated as off-grid (standalone) systems. You need all the following basic components and a lot of things in between:

- ✔ Wind turbine
- ✔ Tower and foundation
- ✔ Transmission wiring and conduit
- ✔ Inverter
- ✔ Charge controller and dump load
- ✔ Battery bank
- ✔ Utility interconnection equipment
- ✔ Possible backup generator

You may wonder why you need the *dump load,* which absorbs surplus energy. If you're thinking that the grid takes all the surplus, "spinning your meter backward" and creating a credit, you're correct. The inverter sells the battery bank down to the set voltage, and the grid absorbs all the surplus energy that the house isn't using. But when the grid goes down, you're back to being an off-grid system, essentially, and you need a diversion controller and dump load to take on the surplus energy generated with most wind generators.

Sizing for a grid-tied system needn't be based on your household's total energy load, though that's a good place to start. (Chapter 6 gives you details on how to calculate your load.) I like to propose to clients that they aim for somewhere near zero energy so that the system produces the same amount of energy the household uses in a year. But sometimes when people see the price tag, they back down somewhat and choose to size by other parameters, such as budget or *zero bill* (a system brings your bill to nothing when you consider energy offset plus incentives). However you size your system, I encourage you to install one that has a reasonable impact, not a token system that has high visibility but generates only a tiny fraction of the electricity of your home.

Because you have battery backup, you also need to choose the size and type of your battery bank. The size depends on how much you want to back up for how long. Backing up a whole house full of appliances for very long on a battery bank is unrealistic. If that's your goal, you should probably purchase a fuel-fired generator. But if you can constrain your backup need either to a small group of efficient appliances or to a short time, a battery bank may be the answer.

The battery sizing math is simple: watts × hours = watt-hours. The battery type decision depends on your budget and your willingness to do maintenance. Sealed batteries are more expensive and don't last as long. Flooded (refillable) batteries require maintenance, but they give longer performance for a lower price. See Chapter 15 for more on this. Because you have batteries in the picture, your system (battery and inverter, at least) voltage will likely be 48 VDC for a whole-house system.

Standing Alone: Off-Grid Systems

Some people don't mean "off-grid" when they say the term — it's just shorthand for living on renewable energy. When I say *off-grid,* I mean systems that have *no* connection to the utility grid.

Many, many times I hear from folks who idealize off-grid living. They think only of the benefits, not the responsibilities. I've lived off-grid for nearly 30 years now and have raised a large family without the benefits and liabilities of the utility grid. I'll tell you straight up that it's not cheap, easy, or dreamlike. It takes time, money, attention, and work.

When you pay your utility bill, you're paying for lots of men and women in suits and coveralls to do all the jobs that a utility does — finance, research and development, generation, business, maintenance, repair, and so on. But if you're off-grid, you're your own utility. You still need to serve all those functions. You may avoid wearing the suit, but you have to do all the jobs, which includes either buying and wearing the coveralls or hiring someone to do that work.

Off-grid living includes some very satisfying parts and some very challenging parts. Knowing that you're responsible for your own electricity and not subject to the vagaries of the grid (with outages) and the steady increase in prices is fantastic. It's wonderful to feel the wind blow and know that it's filling up your battery bank and keeping your lights on and that it's all your doing.

No one has trouble with these good feelings. But successful off-grid systems don't happen because of good feelings; they happen because of the hard work of designing well, installing well, and living well with these systems. This section looks at the advantages and disadvantages of choosing the off-grid option, as well as its components and sizing.

The pros: Independence, flexibility, and being mindful of what you're using

The advantages of being off-grid include the following:

- **Independence:** You don't have to depend on the utility, pay a utility bill, or deal with its policies. You won't experience utility rate increases, outages, or brownouts.

- **Reliability:** Well designed and maintained off-grid systems provide reliable electricity 24/7/365. A properly designed and operated off-grid system has no blackouts.

- **No line extension cost:** For remote properties, the cost of line extension can be very high. Even if you're only a few hundred feet away from the nearest line, extending the lines can cost thousands of dollars, depending on the policies of your local utility. In some cases, line extension can be hundreds of thousands of dollars. At times, extending the line a half mile to your house can cost $60,000 or more. This sort of change can buy a pretty good-sized wind-electric system, or better yet, a hybrid wind/solar-electric system. These situations are the most likely situations for a small wind system to pay for itself. Without the grid for competition and compared to generator usage, a small wind system becomes a very attractive investment.

  In other cases, the grid may be affordably close, but state or utility policies make connecting difficult or uneconomical. Perhaps there's a substantial base fee so that even if you make as much electricity as you use, you're still paying hundreds of dollars a year for the ability to connect to the grid. In this situation, you should weigh your options carefully, because being able to store your surplus on-grid as a credit may indeed be worth the cost.

- **Lower property cost:** Off-grid property is often less expensive than property that already has utility service.

- **Easier to add other sources:** Off-grid systems may be a bit easier than on-grid systems as far as adding other energy sources. These systems are typically lower voltage (generally 48 volts DC for whole-house systems), and therefore adding solar-electric modules can be done more easily and in smaller increments. Because any number of sources can charge a battery bank, there's not an issue of directly matching the sources and inverter(s).

✔ **Energy consciousness:** When you have to make all your own energy with renewables, you think hard about what you really need and about how to make the best use of what you have. This is one of the biggest advantages of off-grid systems in my opinion.

✔ **Emotional perks:** Less tangible advantages are the satisfaction, peace of mind, and the knowledge that you're being a trendsetter by taking full responsibility for your energy generation and use.

The cons: Costs, maintenance, and wastefulness

I'm not a pessimist overall, but I like to be realistic. As a long-time off-grid system owner and manager, I think I clearly see the drawbacks of going off-grid. Among the significant disadvantages are the following:

✔ **Monetary cost:** You won't have a utility bill from the local branch of a multinational energy company or regional electric co-op, but you will still have a cost. And in fact, kilowatt-hour for kilowatt-hour, your cost will very likely be higher than if you were paying a bill to the local utility. Off-grid systems typically deliver electricity that's more expensive than the utility grid, if the grid is available on your site.

Remember: Strictly in monetary terms, small renewable energy systems do not compete well with heavily subsidized utility electricity. However, if you put a true price on utility electricity (which is mostly nonrenewable), including the damage to environment, health, communities, and so on and all the subsidies, renewables usually look like a bargain. For details, see Chapter 10.

✔ **Monitoring and maintenance:** When you go off-grid, the jobs that the utility did for you in the city don't disappear! You have to do all the jobs those guys and gals in suits and coveralls did for you before. Maintaining and operating an off-grid system is a significant undertaking.

Even if you decide to hire someone to design and install the system, you can't hire him or her to live with it. Off the grid, energy becomes a commodity that you have to be aware of, and you need to modify your behavior accordingly (for details, see the sidebar titled "Managing energy for daily life off-grid"). Off-grid systems require regular monitoring because you're dealing with a system that has a limited energy supply. Unless you have an automatically starting generator (which I don't recommend), you need to check your battery bank's state of charge on a regular basis.

To get good life out of a battery bank, you must avoid deep and prolonged discharges of your battery bank and make sure it gets fully charged at least twice a week. Regular battery maintenance is required, and you'll face costly battery replacement every 5 to 10 years (see

Chapter 19 for more on maintenance). Though batteries are one of the most recycled commodities, there's still an environmental cost associated with all that lead.

✔ **Inefficiency:** Batteries aren't just an upfront cost, a maintenance cost, and a replacement cost. They also take an energy toll. In off-grid systems, this is twofold. First, you have to keep your batteries topped off and also periodically overcharge them. This requires some of your wind electricity and probably also some juice from a fuel-fired engine generator at times. In addition, there will be many, many times when you're making wind electricity but have no place to use it, because your batteries are full and your loads are all being supplied. This is an inherent inefficiency and waste in off-grid systems.

Some people have proposed making hydrogen with the excess energy, but this is a terribly inefficient (in the single digits) and expensive way to go. A better strategy is to have what I call *opportunity loads,* such as water heating, an electric vehicle, an electric chainsaw, or other things that you can opt to use when you have surplus. The drawback is that you have to restrain your use of these things when you don't have a surplus.

✔ **Sizing:** With on-grid systems, you can lean back on the grid whenever you're short of wind energy. Not so off-grid. Your system needs to be sized to cover the whole load. This almost always means you'll have another source and often two — solar electricity (see Chapter 11) and a fuel-fired generator are the most common prospects.

✔ **Generator backup:** Unless you're willing to cut back dramatically on your usage, you'll need a backup generator to avoid deeply discharging and damaging your battery bank, even if it's only a few times a year. A backup generator can be costly, dirty, noisy, and a maintenance sink. Fossil-fueled generators are a common failure point in off-grid systems. People who live with generators know that this is not cheap electricity and dream of using the generator less and less as they increase their efficiency and the size of their renewable generation systems.

My advice is to use the grid if it's available at a reasonable cost. Off-grid living is not the idealistic dream that some people portray. It's a big responsibility, with real advantages and disadvantages. Be realistic if you choose this option.

The basic components and configuration sizing

For an off-grid system, you need all these basic components and a lot of things in between:

✔ Wind turbine

✔ Tower and foundation

✔ Transmission wiring and conduit

✔ Inverter

✔ Charge controller and dump load

✔ Battery bank

✔ Backup generator

Off-grid wind-electric systems tend to be the most complicated of all the system configurations, though the configuration may actually look simpler without the grid connection equipment. In a way it is, but designing for a system that has no utility backup means you have a full-time charge controller and dump load and a significantly larger battery bank. These often mean more programming and tweaking of settings — as well as more need for reliably installed and configured systems.

Off-grid system design has to focus heavily upfront on load analysis. Because you must make *all* your electricity, you need to know how much you use and get specific about where you use it (see Chapter 6 for details). The cost also typically inspires you to cut down on this energy usage through efficiency and conservation measures (see Chapter 7 for information).

Managing energy for daily life off-grid

If you've grown up on the grid, you've become accustomed to a more or less unlimited energy source. Okay, so it's not unlimited — if you leave the 1.5 kW heater on indefinitely, your pocketbook will find a limit next time the bill comes. But you're not faced with this daily or weekly, as you are with off-grid systems. Off the grid, energy becomes a commodity that you have to be aware of, and you need to modify your behavior accordingly, one way or another.

In my family's home, we have meters prominently displayed in our kitchen/hallway so all can see what our batteries are doing on a regular basis. I look at the meters probably three dozen times a day or more. And I have a bank of meters right next to my computer so I can tell what's going on with my system whenever I want to be distracted from work.

We regularly have to decide how we'll use energy depending on the supply. When wind and sun are in short supply, either we have to reduce our usage to maintain the health of our batteries, or we have to fire up "the noise" — our propane generator. The kids know that some days it's okay to do laundry and some days it's not. Or that if it's a *not* day, they'll have to run the generator to do major jobs like washing laundry or running the dishwasher or using large shop tools.

Our friends envy us when we talk about "turning on all the lights" when there's a wind storm and the rest of our grid-serviced island is suffering from a utility outage due to trees toppled by the storm. That's the upside. The other side is managing energy when not enough is coming in. Both are part of being off-grid.

Your final kilowatt-hours per day figure is the key number around which to design your system. Combined with your average wind speed (see Chapter 5), daily energy use is the major factor in choosing your turbine size. What if you can't afford a big enough turbine to make most of the energy needed? Or if your wind resource is such that you have long periods without energy input, what do you do?

You must fulfill the balance of your electrical need from other sources, perhaps a solar-electric array and most often a backup fuel-fired generator. Even people with solar and wind resources and generating capacity usually need backup on occasion. If my family lives carefully, we can go through a winter using our backup generator only a dozen times. When we're not willing to live so carefully, the generator runs three or four times that often.

The other critical part of a battery-based system design is sizing the battery bank. This is essentially a question of how long you want to go without wind before you start your generator. It's common to get sucked into thinking that a large battery bank is a good thing, giving you more independence. But in fact it's a costly and wasteful part of your system because in most systems, you already have a backup generator.

Buy a modestly sized battery bank and invest the saved money in a larger wind turbine or solar-electric array. Doing so creates a larger surplus on occasion but generally reduces the engine generator run time. In short, I suggest investing in renewable capacity rather than in lead and acid that you'd have to buy, maintain, and replace.

Batteries have no magic. They make no energy. They only store energy, and they actually waste some of the energy you make, so the smaller your bank, the smaller the waste, up to a point. To make the best use of a system, off-gridders need a bank large enough to carry them through two to three days of no wind.

Another major decision you have to make with off-grid systems is battery voltage. This is an easy decision in most cases. Unless you're on a boat, in an RV, or in a tiny cabin, your battery voltage will be the battery-based system standard of 48 VDC. Gone are the days of using 12 VDC in normal houses — this doesn't make sense from a wire sizing and efficiency standpoint. (Chapter 15 has more details on batteries and their sizing.)

Making Your Decision with Some Considerations in Mind

Think carefully about the three basic system configurations that I describe earlier in this chapter, and make a wise choice for your values and situation. Changing your mind later isn't always easy or inexpensive. Converting an off-grid system to a grid-tied system with battery backup may be very simple if your inverter is grid capable. On the other hand, converting a batteryless grid-tied system to one with battery backup involves a major overhaul of the system wiring, new components, and significant expense.

If you're on-grid, ask yourself these questions:

- ✔ How frequent are utility outages?
- ✔ How do you react during an outage? Are you happy to live with candles for a few hours or days, or do you already own a generator or have a strong need or desire for backup during utility outages?
- ✔ Are you willing to have batteries — and the cost, care, and replacement they entail?

If your property is off-grid, ask yourself these questions:

- ✔ How far away is the utility grid, and what will connecting it cost?
- ✔ Will the grid move closer in the future?
- ✔ Are you likely to connect in the future?
- ✔ Are you willing to live with the responsibility of an off-grid system, including battery maintenance and replacement; generator purchase, use, maintenance, and replacement; and conservation of energy?

All three system configurations can be reliable, long-lasting, and robust. The most important decision is whether you and your values and situations match the system you decide on.

Being on the grid with an off-grid mindset

To me, the best of both worlds is to look at the off-grid community's experience and apply it to your on-grid living. From the off-gridders, discover the meaning of energy analysis (see Chapter 6) and super energy efficiency (see Chapter 7). Find out how to squeeze the most out of a kilowatt-hour, and develop a new awareness of how people use and waste energy in North America today.

Take energy awareness on-grid. On-gridders don't *have* to conserve or be super efficient; you can be as wasteful as you want. But if you take the off-grid awareness and lessons to your on-grid wind-electric system, you'll get the most out of it. If you can become an energy sipper, your system size will shrink, and along with it the cost, environmental impact, and — if you design well — the cost and trouble of upkeep and repair.

So whether you're off-grid or on, I hope you'll take the lessons of off-gridders to heart. Energy awareness and efficiency are the most important things to know. And if the grid's nearby, you can take good advantage of it and change your energy lifestyle by choice, not necessity.

Chapter 10

Calculating the Value of Your Investment

So you've conducted a home energy assessment, increased your home's energy efficiency, determined that you have a viable wind resource, and defined your preferred relationship to the grid. But before you take the leap and convert to wind energy, you need to be clear about what your motivations and goals are and not assume that others' motivations need to be yours. Then you can look at the financial numbers and take that process as far as you choose. Financial payback isn't the be-all and end-all of human life, but if you're considering wind energy for your home, you deserve to understand what you're getting into and how it'll affect your pocketbook.

In the end, deciding which values to apply to your investment in wind energy, and how you feel about it, are up to you. This chapter explains what you need to know before you make your final decision.

Reviewing Your Wind Energy Goals

Without knowing why you want to use wind energy, how can you decide whether wind energy will help you reach your goals or whether a particular system will achieve them?

Analyzing your motivations may seem superfluous to you, but I think it's vital. So take the time to examine your thinking. Be honest with yourself and the people you're talking to about wind electricity. There's no shame in having any of the following motivations, which I introduce in Chapter 2. I'm okay with your wanting a wind generator because you like to see spinning things in the air, you like working in high places, or you like thumbing your nose at the utility. But in the end, it isn't important if I'm comfortable with your motivation. It *is* important that you're clear with yourself about why you're thinking about using wind energy and that you're comfortable with where you're coming from and where you're going.

Saving the Earth

Many people come to wind energy with environmental motivations, myself included. These people are concerned about where energy comes from and how it affects their neighbors and their grandchildren.

Different forms of nonrenewable energy have big environmental impacts. For example

- ✔ **Coal:** The documentary *Kilowatt Ours: A Plan to Re-Energize America* opens with these challenging lines: "What if, every time we flipped a light switch at home, a mountain exploded in West Virginia? Or every time we turned on the air conditioner in summer, a child suffered an asthma attack?" Film producer Jeff Barrie's questions are not theoretical. In fact, it's a very real connection that he makes, because 52 percent of U.S. electricity comes from coal. The coal industry is indeed blowing the tops off mountains in Appalachia while producing air pollution that leads to a variety of lung diseases. And your light switch is connected to this! Every time you switch it on, you're keeping this damaging process going. Roughly speaking, generating 1 kWh of electricity requires a pound of coal.

- ✔ **Oil and gas:** Oil- and gas-fired plants damage the environment at the extraction level as well as when these fuels are burned to generate electricity.

- ✔ **Nuclear:** Nuclear power leaves behind dangerous wastes for thousands of years and risks contamination from this and from reactor accidents.

- ✔ **Hydroelectric:** Even big hydro has serious impacts, flooding communities and ecosystems and restricting the natural passage of everything from tiny fish to kayakers.

If you're interested in wind electricity for environmental reasons, you're far from alone. On the large scale, wind electricity is a big part of the potential solution, and major environmental organizations from the Sierra Club to the Audubon Society have supported wind farms.

If your primary motivation for using wind energy is environmental, do energy analysis and efficiency work before you start thinking about investing in a wind-electric system. This work yields faster and cheaper environmental benefits. (See Chapters 6 and 7 for details.)

In addition, I'd caution you to look carefully not just at the environmental benefits of a wind-electric system but the environmental costs. Everything you buy, build, use, or do has an impact, and installing a wind-electric system is no different.

Consider these facts: With solar-electric (PV) modules, the energy payback is one to four years. That is, in one to four years, depending on the product and site, a PV module generates all the energy it took to manufacture it. From then on, it's actually producing new, clean energy. I've never seen such energy payback analysis on wind-electric systems, and I bet it varies wildly. Freestanding towers, for instance, use large amounts of concrete. One local 165-foot tower has about 70 cubic yards of concrete in the ground. Now that's an environmental impact! I wonder how many years of making 8 to 40 kWh a day it takes to work that off.

My advice is not just to assume that wind-electric systems have environmental benefits while ignoring the liabilities. Well-designed and installed systems can have long-term environmental payback in many cases. You need to be aware of the liabilities and try to minimize them. Building a system with less concrete and steel and with robust components that last a long time will make for the shortest energy payback times. See Chapter 11 for a discussion of the benefits and liabilities of other renewable energy technologies.

Saving money

With wind energy, asking financial questions is reasonable. Many people come to this field wanting to save money. As you find out in the later section "Getting a Grip on the Big Picture," I tend to think this value is overemphasized or at least needs to be balanced by looking at other values. But you may have this as your primary value — you want to save money on your energy bills. Except in some off-grid situations where electricity is very expensive by other means, it's hard to speak honestly about small wind-electric systems as an ideal way to save. I suggest that before you pursue a wind-electric system, you review Chapters 6 and 7, where you can find the primary and highest return activities if your goal is saving money.

If your overriding wind energy goal is to save money, tell the people you're dealing with upfront (such as your suppliers, contractor, and consultants) that that's where you're coming from. And try to find someone who at least understands and can speak to your goals, if not someone who shares your goals. Humans communicate better with others who share their outlook, and you'll have the best experience exploring your options if your guides share your goals. I'd mistrust any salesperson who promises short-term financial gains, and I recommend seeking independent advice.

Fulfilling personal motivations

The category of personal motivations actually includes several, and you can find a bit more discussion of them in Chapter 2. In fact, *all* motivations are personal, but these are less easily defined than environmental and financial motivations. Here are just a few:

- ✔ **Reliability:** Consistent, reliable electricity is one common goal of wind-electric system users. You may want reliability for medical needs, for communications systems (in case you need phone, the Internet, or radio in times of emergency), or simply for comfort and enjoyment.

- ✔ **Technology:** Some people like tinkering with equipment and make no exception for renewable energy systems. They may want to put together a home-built wind generator or rebuild equipment from days gone by. On the other end of the hobby scale are people who like working with the latest gadgets and enjoy working with high-tech airfoils, generators, and electronics.

- ✔ **Trendiness:** Perhaps the largest class of people who aren't environmentally or financially motivated is people excited by what I call *the cool factor*. The marketing departments of everything from clothes companies to car companies speak to this motivation. People like to look good, feel good, and have their neighbors think they're good. The concept is no different with wind generators. Those who live with the wind generators *do* tend to think of themselves as cool, and I have no embarrassment about this. In fact, I think people should play to this motivation and encourage the making of cool wind-electric systems — as long as they can be cool *and* still work well.

If your motivation falls into this category, be clear about it with yourself and the people you're dealing with. If you want reliability, make sure you get it. If you want the latest technology, don't settle for equipment that was designed ten years ago. And if having a system that looks really cool is your top priority, make sure that you and your contractor focus on the aesthetics of the system, not just the functionality.

Analyzing Costs and Incentives

Personally, the financial side of wind-electric systems is not that important to me. I come to wind energy with most of the motivations I've discussed earlier in this chapter, and though the dollars are the weakest of my reasons, even I want to have some idea of what things are going to cost upfront and what benefit I'm going to get. So if financial motivation is fairly (or very) important

to you, you should total your wind-energy costs and incentives and then compare them to what you're currently paying for energy.

But even if saving money isn't one of your motivations, cost may be a limiting
factor on which generator you choose, how big your tower can be, or which
installer you work with. Perhaps you simply want your money to do as much
good as it can. Here's why a financial analysis may still be worthwhile for you:

✔ By viewing finances over the long term — rather than just initial costs —
  you may be more willing to choose a longer-lasting, more productive
  system, which is better for your pocketbook and the planet.

✔ By factoring in discounts and incentives, you may find that you can
  afford to spend more on a better (and cooler) system at the outset.

✔ You may conclude that a wind-electric system is impractical for you and
  that your money may do more good if you use another form of renewable energy or buy green power from the utility (all of which I cover in
  Chapter 11).

Working up a detailed cost/benefit analysis includes looking at not only the
upfront costs but also the ongoing costs. And to compare wind electricity
to your other options, you need to know your present electricity costs and
which incentives are available to you for a wind-electric system. In this section, I explain what you need to do.

Considering installation costs

If you think you can make a decent amount of wind electricity for $10,000,
you have very unrealistic expectations. For a real-world, productive, well-
installed home-sized system, the cost will range somewhere between $20,000
and $200,000. Going cheap with wind energy is one of the worst mistakes you
can make. I can almost guarantee that a system that's cheap to purchase and
install will be very expensive in the end, either through high maintenance
and repair costs or low production (read "expensive electricity").

Ideally, you go through the processes of energy assessment, energy efficiency
work, and resource assessment (see Chapters 6 through 8) with a qualified
dealer. You and the dealer conclude what your need is, what your resource is,
and what your budget is. Then you can determine what the installed system
cost will be. (Chapter 12 explains how to find qualified people to work with.)
Coming up with a specific installation cost is as easy as requesting quotes
from a few local system installers.

How installers respond to your request for a quote is some indication of the quality of their approach. If they churn out a quick quote for a generic system, I recommend running the other way. The best installers begin by asking you about your energy load (see Chapter 6) and continue by talking about energy efficiency (see Chapter 7) before they turn to the question of wind-electric system design. So don't be surprised if they begin to ask you questions about your energy use when you ask what it'll cost — that's a sign of a good dealer.

Skirting around questions about your energy use for a brief moment in your first approach to wind electricity is reasonable. Ask your local installers to tell you the costs of the last five systems they've installed. This will give you a range to work within and a general idea of what you're up against financially.

Adding in maintenance costs

Make no mistake — home-scale wind-electric systems take maintenance. Though I've been involved in or aware of hundreds of small-scale wind installations, I've never seen one that didn't have some issues at some time in its life. And if you don't regularly check and maintain these systems, they'll have more issues, some of them fatal. See Chapter 19 for more on maintenance.

So don't think you can buy a wind-electric system and just walk away. You can just about do this with a batteryless solar-electric system, because the energy collectors (PV modules) are so rugged and simple. Wind generators, however, are a different story. Here's why:

- ✔ Turbines live in a very harsh environment, high above the ground, where they're punished by very strong and variable winds.
- ✔ Turbines are rotating machines that spin, sometimes at very high speeds. Where you have mechanical motion, you'll have wear and tear.

One guideline is to allow 1 percent of the installed cost of a wind-electric system per year for maintenance. Under this rule, for a $60,000 system, you should expect to pay about $600 per year for maintenance. This is a reasonable place to start, but the true cost depends a great deal on the type and height of tower and the quality of the system in the first place. Round up if you have high average wind speeds, because turbines that work harder will probably suffer more damage and require more maintenance and repair.

With smaller systems (perhaps less than $50,000 installed cost), I'd be careful about using this 1 percent rule. If you're hiring out maintenance, I wouldn't expect to pay only $200 per year for a $20,000 system, because the minimum service call fee of a qualified technician may be $200 or more. And if you're in the country with tall trees, your tall tower will mean a higher maintenance bill. By the time you have a climber and a ground person on site with the appropriate gear, a service call for any 120-foot-plus tower may exceed $500.

Paying for your wind-electric system

If you were buying a car, you'd have a number of financing options. Automobiles are part of an established, large industry that's been around for decades. The small wind industry is a baby by comparison, so you won't find well-plowed ground when it comes to financing.

That's not to say that you won't be able to find financing. But it's unlikely that you'll discover a wind-generator company financing your purchase directly. And you may need to dig deeper than conventional financing circles to find a banking institution that will help you get your dream wind-electric system. More and more small and medium-sized banks are discovering that their customers want "green" additions of one sort or another. Finding the right person in these organizations is your task. I recommend getting connected in the renewable energy scene in your area and asking around about what others have done for financing.

Ask your installer (or installers in your neighborhood) what he or she charges for a year of maintenance. Then double it, and if you actually do the maintenance and also have good luck, you may end up with a good number.

Accounting for incentives and discounts

Though the idea goes against my independent streak, many people are looking for help from others to decrease the cost of their wind-electric system. This may come in the form of a grant from the USDA, tax breaks from Uncle Sam (or his nephew in your state), or even incentives from your utility.

The Database of State Incentives for Renewables & Efficiency (DSIRE) can help you locate incentives of all kinds. You can find this very handy Web site at www.dsireusa.org. Click on your state, and you find a complete list of all renewable-energy incentives, with links to detailed data about how they work and how to get them. If you want a more-targeted search, DSIRE also lets you search by technology — you can select "Wind (Residential)" and find state-by-state listings.

Direct subsidies

Subsidies are typically available from two sources — governments and utilities. State, county, and local governments have all been known to offer subsidies for renewable energy systems. And utilities often have their own programs or administer state programs.

You need to factor any direct subsidies, which are new money in your pocket, into your calculations. These generally come in two forms:

- ✔ **Upfront subsidy (buy-down):** An upfront subsidy is often calculated on the rated output of the system. This is a foolish way to think about wind generators, because rated output (peak watts) does not necessarily correlate well with actual production (kilowatt-hours) (see Chapter 5 for details).

  Under this sort of subsidy, a user may be paid "$4 per rated watt," so a "1,000-watt" wind turbine would be granted a $4,000 incentive. Note that this incentivizes systems with high ratings, but that turbine could be installed and maintained poorly and generate very few kilowatt-hours.

- ✔ **Percentage of total system cost:** This is another questionable plan because the cost of a system also does not correlate with energy production. Though obviously you would hope to get more kilowatt-hours if you invest more dollars, the more important factors are whether your system is well designed, sited, and installed.

  You could spend $100,000 on a system, get a $50,000 grant, and install the turbine on top of your garage where there's almost no wind. It may look like you got a good deal, but if your goal is lots of kilowatt-hours, you didn't. And the taxpayers or ratepayers who funded the grant scheme really got a bad deal.

To make sound choices, you shouldn't get starry-eyed about high incentives and low system cost. Keep your eye on the prize — renewable kWh. You should account for subsidies when calculating how much a system will cost but make sure other factors take priority in your choice of turbine, design and installation decisions, and so on. A low price on a poorly producing system is no bargain.

If the goal is to encourage renewable energy, subsidies would be more effective if they were to encourage energy, not peak power or high cost. In other words, subsidies would be better paid on a per kilowatt-hour basis, after the energy is generated. Then you don't have to argue about power ratings or percentage of cost, and you put the responsibility and incentive to install systems well squarely where it belongs, in the owners' hands.

Tax incentives

Tax incentives for wind-electric systems can be substantial. At present in the United States, there's a 30 percent federal tax credit for wind-electric systems. This requires that you have a tax liability (can be rolled over two years), but if you do, you can pay for 30 percent of the installed system cost instead of paying Uncle Sam. Which would you rather do?

Many states also exempt renewable energy systems from sales tax, which can cut the cost another 8 percent or more in some cases. And businesses can use depreciation (sometimes accelerated) to take even more off their tax bill.

Net metering

One subcategory of incentives is *net metering*. About 40 U.S. states have net metering laws, which essentially require the utilities to buy your electricity at the same rate you pay them for electricity. It's called *net* metering because you can typically get this attractive rate only up to the point where you make more (in a month, year, or other period) than you use. But up to the level of your usage, you can "spin the meter backward," creating a credit to use in less windy times. (If you make more than you use, the utility may not pay you at all; or in some cases, you receive wholesale or "avoided cost," neither of which is a lot.)

I said that the rate was attractive, and in fact it is. I know of no other industry where a business is mandated to pay the suppliers at retail rates. If this portion of the utility's business were more than a fraction, it wouldn't be a sustainable strategy. But in this infancy of renewable energy, the idea is that states can encourage this new technology in this way.

Considering the cost of electricity you don't have to buy

If you're going to look at either *payback* (how long it takes for your investment to pay itself off) or return on investment, you need to know what you're comparing your wind electricity to. Your financial goal with wind electricity is actually fairly easy to define: You want to pay less for electricity. After you've done your efficiency work (see Chapter 7), paying less means paying less per kilowatt-hour (kWh).

You compare costs from the wind-electric system to the cost per kilowatt-hour of electricity from your local utility. This number varies widely depending on your region, locality, and the type of utility you're dealing with. States with high energy costs, such as Alaska, Hawaii, and California, may have rates as high as 40 cents per kilowatt-hour, whereas other places, such as Idaho, may see rates below 3 cents per kilowatt-hour. (Of course, different areas within a given state have costs above and below the average.) The cost depends on the source of energy and the form of delivery organization. See Figure 10-1 for average figures by state and region; check your rates locally.

 To find out your true cost per kilowatt-hour, take the total dollar amount of your utility bill and divide it by the total number of kilowatt-hours used in that billing period. This provides you with a baseline with which to compare the projections you'll make of the cost per kilowatt-hour for wind electricity on your site (see the later section "Crunching all the numbers" for details).

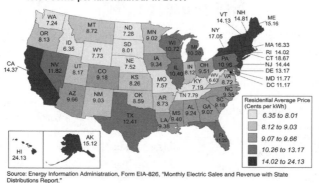

The U.S. average residential retail price of electricity was 10.64 cents per kilowatthour in 2007.

Figure 10-1: The average cost of electricity per kilowatt-hour in the United States.

Source: Energy Information Administration, Form EIA-826, "Monthly Electric Sales and Revenue with State Distributions Report."

Crunching all the numbers

Looking at cost per kilowatt-hours provides a basic and sensible way to look at the dollars without breaking too much of an accounting sweat. (I provide a simple example here, because I'm not your accountant or even my own. You money nerds can expand on what I've started, or seek some online sources of sophisticated financial analysis, or hire consultants.) Here's how the number-crunching works:

1. **Determine your total out-of-pocket cost for the wind-electric system.**

   First find the wind-electric system's total installed cost. For instance, suppose that Bill and Mary plan to purchase a substantial wind-electric system for their rural home, and the total installed cost, including design, prep work, installation, and materials, will be $82,000. In their case, they've invested well in a wind generator with a solid track record, and their site is very accessible to their local contractor.

   Then subtract tax incentives and direct subsidies based on the cost of the machine. Bill and Mary can take full advantage of the 30 percent federal tax credit. In addition, their system has no sales tax. Take their $82,000 starting cost and subtract the 30 percent tax credit ($24,600). This makes their out-of-pocket cost $57,400:

   $$\$82,000 - (30\% \times \$82,000) = \$57,400$$

2. **Add the cost of maintenance over the life of the machine to your total out-of-pocket cost.**

   Earlier in "Maintenance costs," I note that one guideline is to allow 1 percent of the installed cost of a wind-electric system (before discounts) per year for maintenance.

In the example, suppose the contractor actually develops a long-term service contract and is going to charge Bill and Mary $750 for two inspections per year, with routine maintenance included. Bill and Mary sign a contract with the company to do it for the next 10 years at that price (a total of $7,500), with an option to extend it another 10 years with a maximum of a 20 percent increase ($900 per year) for the second 10 years (a total of $9,000). (This is actually hard to find in the real world, but it simplifies the example.)

The lifetime for this system is 20 years. Add the maintenance costs, which at $7,500 + $9,000 amounts to $16,500 for 20 years of maintenance. Now the total outlay (assuming no acts of God, major failures, upgrades, and so on) is $73,900:

$$\$57,400 + (\$750 \times 10 \text{ years}) + (\$900 \times 10 \text{ years}) = \$73,900$$

3. **Figure out the machine's total predicted kilowatt-hours over its lifetime.**

Multiply the machine's yearly predicted kilowatt-hour production by the lifetime of the machine. Bill and Mary's wind resource is reasonably good at 11 mph, and the wind generator manufacturer predicts that it will produce about 18 megawatt-hours (MWh) per year — 18,000 kWh. That's 360,000 kWh in the machine's 20-year lifetime:

$$18,000 \text{ kWh/year} \times 20 \text{ years} = 360,000 \text{ kWh}$$

4. **Calculate the wind-electric system's cost per kilowatt-hours.**

Divide the system's cost, including maintenance (from Step 2), by the number of kilowatt-hours the machine produces over its lifetime (from Step 3). For Bill and Mary's system, dividing $73,900 by 360,000 equals about 21 cents per kilowatt-hour:

$$\$73,900 \div 360,000 \text{ kWh} = \$0.21/\text{kWh}$$

At this point, you can subtract any subsidies based on kilowatt-hours produced. Suppose a state incentive (which expires in 7 years but will likely be reinstated) pays 12 cents per kilowatt-hour generated. This brings Bill and Mary's cost per kilowatt-hour down to only 9 cents if the machine performs as advertised, has no major problems, and the current incentive climate stays the same:

$$\$0.21/\text{kWh} - \$0.12/\text{kWh} = \$0.09/\text{kWh}$$

The final cost per kilowatt-hour gives you a good basis for comparing different wind-electric systems. You can also compare this number to utility rates.

If you're interested in figuring out how much money you stand to save (or lose) by using wind power instead of buying from the utility, you can use the cost per kilowatt-hour to calculate simple return on investment. Suppose Bill and Mary live in a rural county in the American West and their utility rate is a rather high 32 cents per kilowatt-hour. The wind-electric system's 9 cents is a much better deal than the 32 cents per kilowatt-hour that Bill and Mary currently pay. In fact, Bill and Mary stand to gain 23 cents per kilowatt-hour generated, which over the course of 20 years will mean around $82,800, or $4,140 per year. (And that assumes that utility rates won't go up.)

To find the *simple return on investment,* divide the expected yearly gain by the total amount of money invested (from Step 2). Investing about $73,900 will give Bill and Mary a return of approximately $4,140 per year ($82,800 ÷ 20 years), or about a 5.6 percent simple return — not bad!

$$\$4,140 \div \$73,900 \approx 5.6\%$$

And not bad at all if you have motivations other than financial, because that's all gravy on top. More gravy will come if the system lasts longer than 20 years, which is quite possible if Bill and Mary invest in a quality system and maintain it well.

Getting a Grip on the Big Picture

The example in the earlier section "Crunching all the numbers" is pretty rosy. All the factors play in favor of Bill and Mary. They have a high utility rate. They have significant financial incentives. And they have a good wind resource. This combination of factors isn't rare, but it's not all that common, either. And it takes exactly this sort of combination to make wind electricity pencil out on a purely financial basis. So if you don't have high utility rates, generous incentives, and a high average wind speed, should you give up on a wind-electric system? Well, that depends on your motivations.

I maintain that very few decisions people make in life are primarily financial decisions. When you buy a house for your family, you don't just look for the cheapest or the one that will give you the best profit when you sell it. You want it to be the right location, style, size, condition, and even the right color. Sure, you're concerned about the price, but it's just one consideration, and the return is often prioritized far behind other more personal factors. Energy should be no different.

With noninvestment purchases, consumers rarely ask about *payback* — how long they'll have to wait for a purchase to pay for itself. Even with things like increased insulation, people often don't calculate the financial return. And they certainly don't do it on things like food, furniture, and clothing. This isn't because there's never a return but because there's a bigger picture. This bigger picture has positive and negative values, and acknowledging them can help people make better decisions for themselves and their communities.

One key piece of finding the true value of energy is understanding that the price you pay at the pump and at the kilowatt-hour meter doesn't reflect the true cost or the true impact of these energy choices. Here are some factors that often put the market price of nonrenewable energy lower than it should be:

- ✔ Direct subsidies to energy companies
- ✔ Tax breaks compared to other energy technologies
- ✔ Government-granted monopolies on resource extraction and distribution
- ✔ Limited liability (especially nuclear energy)
- ✔ Military support for resource extraction and transport
- ✔ Socialized environmental and health costs

Looking at the return on energy efficiency

In "Crunching all the numbers," I paint a very rosy example in which Bill and Mary receive a 5.5 percent return on their investment in a wind-electric system. The right conditions collide to make this possible. But that doesn't mean that I'd suggest that Bill and Mary jump into that investment before doing energy efficiency work.

A simple example is in order. Using one light bulb in their house, examine the simple return on switching to efficient lighting. Say that Bill and Mary have a 75-watt incandescent light in their den, and they use it for 6 hours per day. To find out how much energy they use, multiply watts by hours. You can easily determine that this light uses 450 watt-hours (Wh) per day and therefore about 164 kWh per year. With their utility cost of 32 cents per kWh, this means that Bill and Mary spend about $52.50 to operate this light.

The incandescent light runs for about 2,200 hours per year and therefore needs to be replaced twice every year. At a cost of 75 cents each, that's $1.50 per year, for a total capital and energy cost of around $54 per year for the incandescent light.

A 20-watt compact fluorescent bulb gives about the same amount of light (after it warms up), and costs perhaps $6 to purchase (you can find lower quality and rebated models for less). It will run about 10,000 hours, so say it'll last Bill and Mary four years. At 20 watts, the 2,200 hours mean that the light uses 44 kWh per year, at a cost of $14 (32 cents for each kilowatt-hour). So dividing the bulb cost over the four years, Bill and Mary are going to pay approximately $15.50 per year for the same amount of light, saving $38.50 per year for four years, or $154.

If I have my math right, this gives a 642 percent simple return per year. This makes the return on a wind-electric system look pretty pitiful. A 5.6 percent return on the $6 would be a little less than 34 cents per year, as opposed to $38.50 per year. Where would you rather put your money?

Apply this same logic to other energy efficiency measures, and you may not find quite the high level of return, but you'll find a better return than investing in a wind-electric system. Increasing the insulation in your home, buying an energy-efficient refrigerator, reducing phantom loads — all these measures are probably a better bet than almost any generating system if your goals are financial or environmental. Chapter 7 has details on increasing your home's energy efficiency.

In the end, you don't know the true cost of a coal-made kilowatt-hour, nor the true cost of a gallon of gas. And having poor pricing leads to poor decisions, including deciding that wind electricity is "too expensive." As conscious human beings, individuals can develop a gut feel for the true value and impact and then make better decisions.

In a recent radio interview, I was asked whether I thought renewable energy was "viable." My response was to ask whether blowing off the tops of mountains so you can have coal is viable. When you tally up the damage done and the true cost and impact of nonrenewable technologies, they don't look all that viable to me. Renewables, on the other hand, look like a slam-dunk. Who wouldn't want an energy technology that has infrastructure and operations costs but no fuel costs? Can you imagine the delight of an energy business-person who has spent his or her life generating electricity with coal and oil and then discovering wind power? Wow — no fuel cost!

The situation is similarly delightful on the home-scale. People too often compare small-scale wind with large-scale nonrenewables. A more instructive approach may be to compare small-scale wind with trying to run your house with a diesel generator. Imagine having to pay for the fuel and having to deal with the noise, dirt, and air pollution of a diesel generator in your garage. Then you can see the value of investing in a clean and local technology.

Compare the impact of any technology, because that's part of the full value of any product. You can't just look at the dollar cost and think you're seeing the true value. You can buy a cheap but dangerous car or a cheap but unhealthy burger. Looking at the full value of wind-electricity can make it easier to make a commitment to a clean technology and feel good about your investment decision. To do this, you need to look at more than just financial motivations and goals. See Chapter 2 for a check on your motivations.

And if your design is not penciling out financially, you can also take a second look at your site and system design — a taller tower, a bigger rotor, or a different system configuration may look a little greener for the environment and for your pocketbook. Or perhaps you can find a more suitable renewable-energy option in Chapter 11.

Chapter 11

If Not Wind, Then What? Other Options for Green Energy

*W*ind electricity isn't for everyone. You must have the right conditions and determination to make it work. What's truly discouraging is seeing people overlook, ignore, or bend these qualifications and later become disappointed in the performance, reliability, and longevity of the system they end up with. Here's what you need:

▸ **A reasonable wind resource:** Average wind speeds of 9 to 12 mph on-grid and at least 6 or 7 mph off-grid. See Chapter 8 for details on wind resources and Chapter 9 for information on relationships to the grid.

▸ **An energy-efficient home:** Although this isn't essential, investing in what may be a fairly expensive electricity option doesn't make much sense if you're wasting a lot downstream. Check out Chapters 6 and 7 for more.

▸ **Space to install a tall tower:** You need to be able to get the wind generator at least 30 feet above anything within 500 feet — higher is better. Head to Chapters 5 and 14 for more information.

▸ **Permission to install the system:** This is much more difficult in urban and suburban locations than in the countryside. See Chapter 2.

▸ **A mindset and financial situation that allow you to invest in high-quality equipment and maintain it well:** This is not a cheap or easy ride. Look to Chapters 2 and 10.

Take a careful look at your situation, motivations, budget, and abilities. If you don't fit into the preceding profile, are you out of the running for reaching your goals of a cleaner environment, lower cost electricity, reliability, or being cool? No. As you find out in this chapter, you have many alternative options to reach your goals. And although they may not be quite as fun and challenging, most of them may actually be cheaper and more effective at reaching your goals than wind electricity.

If your goal is environmental or financial, *nothing* will move you faster toward your goals than making energy efficiency changes in your home design, construction, and use. Flip to Chapters 6 and 7 for details.

Here Comes the Sun: Solar Electricity

The blunt truth is that when I'm talking with a client, solar electricity is much higher on my list of suggestions than wind electricity. There — I've said it! Wind generator salespeople can now claim that I'm against wind. But that's not true at all. I'm only against wind-electric systems that don't perform well, disappointing the users and giving the industry a black eye.

And even though I'm an author, educator, and consultant in the small-wind industry, I don't put promoting wind ahead of helping people reach their goals. Rarely do people approach me with the sole goal of having a spinning thing above their homes. Normal folks want cheaper, cleaner, more reliable electricity, and they think that a wind-electric system can reach those goals. Although on the utility scale, wind energy is more cost effective than solar electricity, that isn't often the case at the home scale. Very, very often, solar electricity can reach those goals more effectively and more reliably.

In this section, I describe the simplicity and reliability of a solar-electric system and show you how to compare solar-electric system costs to wind-electric system costs so you can determine which one is the better deal for your situation.

Many local and regional dealers can talk to you about solar electricity and help you with a site assessment and system design. Find reputable dealers in your area by contacting the local solar energy group, using the Yellow Pages, or cruising the installer directory of *Home Power* magazine or their online directory at www.homepower.com. And for more information on solar power, check out *Solar Power Your Home For Dummies,* by Rik DeGunther (Wiley).

Looking at a solar-electric system's longevity, reliability, and simplicity

Photovoltaic (PV) modules are remarkable devices. I challenge you to find another consumer product with a 25-year warranty. If your PV module isn't producing 80 percent of its rated output (maximum output in ideal conditions) 25 years after you buy it, any of the major manufacturers will make you whole, providing module capacity to bring your array back up to spec. All the major manufacturers offer this warranty.

And manufacturers typically expect a product to last twice as long as the warranty period. People have experienced this in real life with crystalline PV modules. Imagine buying a product that can give you 50 years of electricity at a fixed upfront cost.

Now imagine that the product will do it reliably, giving you virtually no trouble and little maintenance. In my 25 years of using wind and solar electricity, my PV modules have just sat on the roof doing their job. I wash them a few times a year, but other than that, I've had no involvement. In the same time period, I've had multiple wind generator failures, and my machines have all needed regular maintenance.

Batteryless grid-tied PV systems are even simpler than batteryless grid-tied wind-electric systems (see Chapter 9). The major components, as Figure 11-1 shows, are few:

- ✔ PV array
- ✔ Racking for the PV array
- ✔ Wiring and disconnects
- ✔ Batteryless inverter
- ✔ AC distribution panel
- ✔ Utility interconnection equipment and metering

PV systems just sit there and make electricity whenever they get sun. The modules are electronic devices with no moving parts, typically with tempered glass fronts that are very tough. With state and utility policies of *net metering* (which requires that the utility pay you at the same rate they charge you), you can use the utility grid as a huge battery, storing your surplus as a credit against future electricity usage.

Another bonus: Besides being simpler and more reliable, solar-electric systems are more modular than wind-electric systems. Expanding as your interest or budget allows is relatively easy (though some planning ahead is advisable if you think growth will be in your future).

In addition, many, many more people have a good solar site than have a good wind site. PVs can go on roofs, poles, and ground mounts in the city, suburbs, or wherever. Good solar sites have clear access to full sun all day, or at least for the peak hours of 9 a.m. to 3 p.m. Though not all sites have good exposure, I'd say that for every good wind site I look at, I probably look at 40 or 50 good solar sites. See NREL's solar resource maps at www.nrel.gov/gis/solar.html to get an idea of your solar resource.

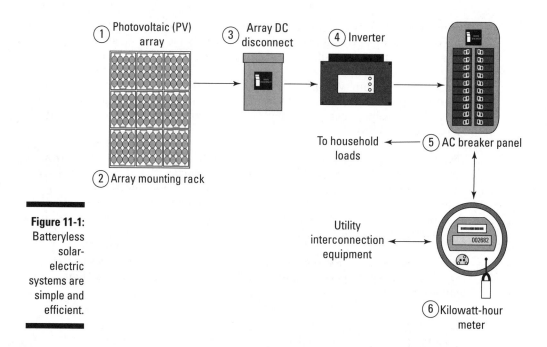

Figure 11-1:
Batteryless
solar-
electric
systems are
simple and
efficient.

Comparing solar- and wind-electric costs

Before you jump to the conclusion that PV is expensive, remember that you have to look not only at the initial investment but also at the life-cycle cost and cost of energy (see Chapter 10). Compared to wind, PV has significantly lower maintenance costs, negligible failure rates, and typically lasts longer. Make sure you compare apples to apples before you take issue with the cost.

Consider the example of Bill and Mary that I introduce in Chapter 10; with their $74,000 after-tax investment, they yield 360 MWh over the 20-year life-time of their wind-electric system. If you take the same $74,000, you may reasonably expect to get at least a 9 kW rated PV system with a 25-year module warranty; in an average area (with 4 peak sun hours), this would yield about 25 kWh per day, or 365 MWh over the 40-year life of the system. So wind is looking pretty good in this case, giving about the same amount of overall energy for the invested money over a shorter time span.

What isn't factored in is the replacement or repair cost after 20 years or the reliability of the two systems. Making this sort of comparison is a bit treacherous because solar-electric systems almost always live up to this best-case scenario, whereas wind-electric systems most often do not.

Also remember that Bill and Mary have an excellent wind resource. If instead of the 11 mph resource, they had 8 mph, their total production over 20 years would shrink to 150 MWh. Over 40 years, then, a wind system would produce only 300 MWh, meaning that a solar-electric system producing 365 MWh would be more worthwhile.

Understanding the cons of PV arrays

Do solar-electric systems sound too good to be true? Well, they are very good, but many people can't afford the upfront cost. This is their primary drawback, and as the industry becomes more mature, more financing options are popping up. Comparing a PV to a car is instructive. Both are large investments for the typical homeowner, whether you're buying an economy car or a luxury car, a small PV system or a large one. If everyone in the United States had to pay cash for a car, you wouldn't see very much traffic congestion.

The only other drawback of a well-designed and installed PV system is that it doesn't move! I've never heard of a family moving their dinner out onto the deck so that they could watch their PV array. Only tracked PV arrays have any motion, and that's very, very slow, just a tad more entertaining than watching paint dry. A few innovative dealers are selling artistic PV arrays — for instance, arrays constructed in the shape of a flower. But they can't compete with a wind generator for live action and interest.

The Waterworks: Hydroelectricity

Water power is dynamic, though the spinning parts are hidden away. What's more exciting about hydropower is its 24-hour availability and the density of the energy. Wind is very thin, so you need big rotors to get much out of it. Water is about 800 times denser, so a typical home hydro "rotor" — actually called a *runner* — is measured in inches instead of feet. Read on to get some basics on how to tap into this reliable and powerful resource.

Because there are so few good hydro sites, there are also few good hydro dealers and resources. In North America, there are basically four turbine manufacturers and only a handful of installers who've done more than a dozen systems. Seek out these experienced people, even if they're not in your area. Track down the excellent hydro articles in the back issues of *Home Power* magazine. Take one of the hydro workshops offered by renewable energy education organizations.

Tapping the resource (if you have it)

Many people have good solar energy sites. Fewer have good wind sites. Very few people have good sites for hydroelectricity. Think about it: How many of your friends and relations have a good, sunny yard? How many have good exposure to wind and the possibility of putting up a tall tower? And how many have a stream falling down the hill behind their house? In the rare case where all three resources — sun, wind, and falling water — are abundantly and conveniently available on site, hydro is likely the one system to look at first. Why?

Home-scale hydro systems are low impact and low maintenance. You take some water out of a stream for a time, put it into a pipe, and run it downhill. The turbine is small — just a few square feet at most — and powerful enough to run even wasteful homes if you have the resource. The components are few, and many are similar to those in wind and solar-electric systems.

Hydroelectric systems include an intake to get the water out of the stream, a pipeline, a turbine, controls, transmission, and distribution. (Figure 11-2 shows the basic components.) Costs for these systems depend largely on the size and length of the pipe, but for the energy delivered, they're often less expensive than solar-electric and wind-electric systems.

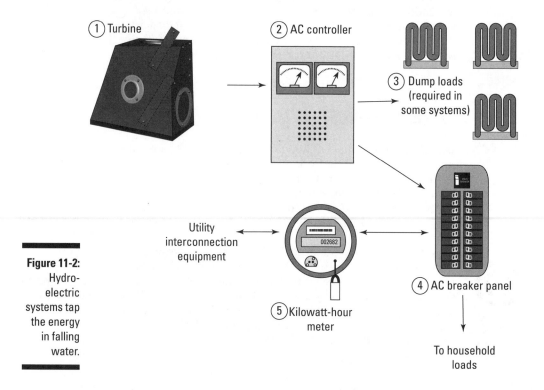

Figure 11-2:
Hydro-
electric
systems tap
the energy
in falling
water.

① Turbine

② AC controller

③ Dump loads
(required in
some systems)

④ AC breaker panel

⑤ Kilowatt-hour
meter

Utility
interconnection
equipment

002682

To household
loads

So what makes a good resource? To tap the energy in water, you need two things — head and flow:

- *Head* is the vertical drop from the point where you take water out of a stream to the point where it hits your hydroelectric turbine. It's typically measured in feet or in pounds per square inch (psi). These two measures are actually directly convertible — every 2.31 feet of vertical drop is equal to 1 psi.

- Flow for home systems is typically measured in gallons per minute (gpm).

You need plenty of head or flow to make a significant amount of energy. A simple formula says that

Head (in feet) × flow (in gpm) ÷ 12 = output (in watts)

So a site with 140 feet of head and 25 gpm would yield 290 W, or 7 kWh per day. The inverse — 25 feet of head and 140 gpm — would give you about the same power and energy. If you don't have much head, you need lots of flow to make much energy; if you don't have much flow, you need lots of head.

Some hydro systems require very long and expensive pipelines. Others run into bureaucratic quagmires that make them impossible to develop. Local flora and fauna use water, too, so you have to limit how much water you take and consider how the system extracts and returns water so the system doesn't impede fish and other residents from traveling where they want. But overall, these systems have a very low impact per kilowatt-hour and require only moderate maintenance, which is on the ground, not 100-plus feet in the air. In most cases, hydro is your best resource to tap if you have it and if the distances are not great.

It's too bad that best-to-tap coincides with fewest-people-have-it. But keep your eyes peeled; you may find a powerful resource that you weren't aware of. If you're shopping for property, look for land with a stream. And don't overlook even the smallest resource — measure the head and flow of the water and do the math. I know of homes running half the year on seasonal runoff from drainage ditches. How little water you need to do something significant, if you have the head, may surprise you.

Enjoying constant electricity

One of the major benefits of hydro electricity is that it's pretty constant. Wind comes and goes. The sun shines only in the day, and trees and clouds may shade your array. Though some hydro systems vary seasonally, most are fairly constant within the seasons that they run — at least they're much more constant than wind- and sun-powered systems. This means a few things:

- ✔ Your energy output is predictable, not only in terms of how much but also when it will be available.

- ✔ If you have batteries, your battery bank can be much smaller, because you know you'll always have this predictable input.

- ✔ Running larger systems without batteries is possible, because you have enough wattage to run all loads at once in backup or off-grid modes.

Looking at the cons of hydro power

Although hydro systems are not magic, they don't have many drawbacks, either. They do require some maintenance, though if the systems are well designed, this may involve no more than adjusting nozzles several times a year and replacing bearings every few years. Poorly designed systems can require cleaning the intake on a regular basis.

The biggest problem with hydro systems is often surfing the bureaucracy. Although there's valid concern about preserving water, animal, and plant resources, hydro systems can be designed to be very low impact. In fact, I'd venture that the swarm of bureaucrats you may have to plow through to get official permission are doing much more damage with their offices, cars, and lifestyles than you'll likely do with your hydro system.

Forget about rooftop hydro

I often hear from folks who want to tap the energy in their downspouts. Let go of this idea now — there's hardly any energy there. For instance, a major Seattle corporation once asked me to study the potential of the rain coming off their roof. At 40,000 square feet and 80 feet high, the roof looked like a big collector. But when I crunched the numbers using Seattle's rainfall, I ended up concluding that the system would make about 1 kWh per day, which costs 8 or 10 cents.

A good way to visualize why rooftop hydro won't amount to anything is to picture how big the collector is for a typical hydro system. For small utility-scale systems, designers look at the collector area in square miles — it's the watershed of the stream or river you tap. Even small streams have collectors that are several square miles minimum. At 28 million square feet per square mile, your perhaps 2,000-square-foot roof has very little collector area.

My late friend Don Kulha and I once calculated and theorized that we could put teensy turbines in our downspouts and they might run an LED indicator that would tell us how hard it was raining. This is the level of the resource coming off your roof.

Hydro systems also have less subtle bragging rights. They fail miserably in being a visible sign of your greenness — you often have to take guests off your manicured lawn and down into a canyon to show off your energy toys.

Heating Things Up with Solar Thermal Applications

Some clean ways to use renewable energy are actually more cost effective than any electricity technology. One option is *solar thermal energy* — using energy from the sun (rather than electricity or gas) to heat space and water.

Using solar-electric modules to make electricity and then using that electricity to heat space or water is an indirect way to do the job, and that method isn't cost-effective in most cases. The same is true of wind on many sites (though better sites can at least use surplus energy for heating, if not for their main heat source). Using the sun to heat space or water more directly makes sense. In this section, I describe three solar thermal applications: passive solar design, solar hot air systems, and solar hot water systems.

 Look at efficiency (as I describe in Chapter 7) before you try thermal applications of solar energy. Make sure your home, water heater, pipes, and ducts are well sealed and insulated. Get good windows and doors, think about how you're losing heat, and address the issues. Only after this work should you begin to look at direct use of solar energy for heating applications.

You can get solar thermal systems, especially domestic solar hot water systems, from many solar- and wind-electric contractors. Passive solar design expertise is more often found with green architects and building designers. Active space heating with sunshine is a more complex discipline, so look for experienced designers and engineers in the renewable energy community.

Warming the house

Home heating is generally the largest energy load of a home anywhere in the northern half of North America. Finding a way to use renewable resources for this job can make your budget and your home more comfortable.

The simplest and most cost-effective method is passive solar design — building a house that *is* a solar collector. If that's not an option (perhaps your neighbor's house or plantings crowd your south walls), solar hot air or solar hot water can be options to consider. I discuss all three options in this section.

Passive solar design

Passive solar design means using windows, *thermal masses* (substances that absorb and release energy), and similar elements to tap directly into the sun's energy, with no need for special equipment. For space heating, passive solar design can work as the main heat source or a supplementary source, depending on the site and climate. If you're trying to be a *zero-energy* home (meaning you make all of your home's energy on-site), using passive solar design reduces the size of the wind-electric or other renewable energy systems that you need to power your home. And houses implementing these designs don't need to cost a lot more than a conventional house, especially if you're starting at the design stage. Retrofitting for passive solar gain is tougher but not impossible.

Figure 11-3 shows a basic passive solar home design. Passive solar homes include the following:

- ✔ **A calculated percentage of glazing (glass) on the south side:** This glazing is designed to make a good balance between solar gain during the day and heat loss at night. The other three sides have modest amounts of glass, with the least in the north (where you get almost no sun) and west (where you overheat if you overglaze, because the home will already be warm by the time the sun hits there).

- ✔ **Thermal mass to capture the sun's heat:** Thermal mass absorbs solar energy during the day and releases it gradually at night. Thermal mass is typically in masonry floors or walls, but it can be in the form of water or even high-tech phase-change materials. The amount of thermal mass is balanced with the home's size and the amount of glazing.

- ✔ **Super insulation, to hold in the solar energy:** This is energy efficiency at its best. Why have to collect or generate more energy when you can just better care for what you have, not allowing it to escape so quickly?

- ✔ **Properly designed roof overhangs:** These overhangs allow the winter sun deep into the house to reach the most thermal mass while preventing the summer sun from entering. The ideal is to have *no* sun enter through the windows in the height of summer and as much as possible enter in the depths of winter.

More and more building designers are discovering the benefits of incorporating passive solar design principles into their client's homes. These homes may look normal overall — not like the overglazed and poorly designed solar homes (really solar ovens) of the 1970s. But they aren't normal in performance. They use an abundant natural resource directly and effectively, avoiding the need for generating more energy on-site.

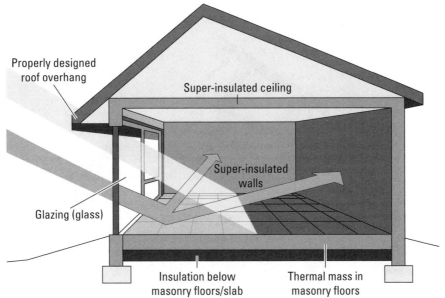

Figure 11-3:
Passive solar home design uses the sun's energy directly.

Properly designed roof overhang

Super-insulated ceiling

Super-insulated walls

Glazing (glass)

Insulation below masonry floors/slab

Thermal mass in masonry floors

Solar hot air systems

If you're starting a building from scratch and have good solar exposure on your south wall, passive solar design (which I describe in the preceding section) makes more sense than a solar hot air system. And if you're retrofitting an existing building, passive solar is still your first, best option. But sometimes, the remodeling costs or the home design make it difficult to take advantage of the sun directly. And often in urban and suburban environments, neighbors' trees and homes severely or completely shade the south walls of houses. If this is the case and your roof has clear exposure to the sun, solar hot air may be a space-heating option worth considering.

Solar hot air systems use collectors that look similar to flat-plate solar hot water collectors (the next section has details on solar hot water systems) — large, glass-fronted boxes about the size of a sheet of plywood and several inches thick. But instead of circulating water in the glass-fronted insulated boxes, air is blown through them. Ducts run down from the roof-mounted collector, pulling cold air from the room and blowing solar-heated air back down into the house. A thermostat controls the blower, turning it on only when the temperature in the collector is higher than in the house. See Figure 11-4 to see how simple this system can be.

Minimize duct length and complexity in a solar hot air system. The best scenario involves collectors very near the rooms to be heated, with short, straight ducts sized large enough to carry the load easily.

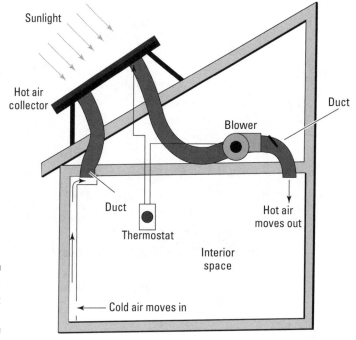

Figure 11-4:
A solar hot
air system.

Solar hot air systems are simpler and therefore less expensive than solar hot water systems. On the other hand, air is a much less effective heat transfer medium, because you need 3,000 times as much of it to carry equivalent heat to the same volume of water. This doesn't mean the solar hot air systems aren't effective, only that they must be well designed.

Solar hot air system costs vary widely depending on home size and climate. In general, they're less expensive than solar-hot-water-to-radiant-floor systems (which I mention in the next section) because they have simpler delivery and control mechanisms.

Solar hot water for space heating

In the right circumstances, you can use sun-heated water for space heating. In this method, the system delivers the heat through radiant floors or base-boards. Again, passive solar design is the simplest and most cost effective way to heat with the sun. But solar hot water can work and work well if you have a good solar resource in the cold season.

If you live in a climate that's cloudy, rainy, or snowy when you most need heat, trying to heat your water and your space with a solar hot water system may not be worth the cost and complication. And because a space heating system needs to be much larger than a domestic water heating system (which I discuss in the next section), you'll need a method of dealing with the surplus you'll inevitably have in the summer. Hot tub time?

Getting into hot water: Solar hot water systems

Typical American homes may use 10 kWh or more per day to heat their domestic water. I've been monitoring some homes with and without solar hot water systems, and that's the shocking truth. This can be a major portion of even a wasteful home's electricity usage and maybe even as much as half of an efficient home's non-space heat energy. So looking into capturing solar energy to do this job is well worth considering.

Even in the modest solar resource where I live in western Washington, solar hot water systems can provide most or all of a home's hot water for about half the year and about half the hot water for the other half of the year. My early tests are showing that they have the potential to cut that 10 kWh a day down to perhaps 2 to 3 kWh per day.

Solar hot water systems incorporate a collector, piping, often a heat exchanger, a tank, and controls and pumps. The collectors may be *flat-plate,* with an insulated box with copper pipes inside, or *evacuated tube*, with high-tech tubes that have a heat-exchange fluid inside. I live with both, and they both work well. Figure 11-5 shows a basic solar hot water design.

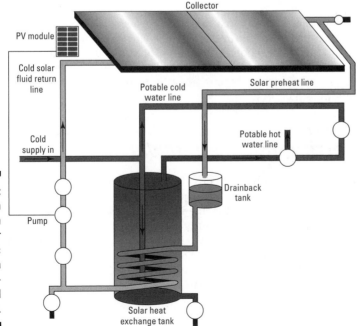

Figure 11-5: The sun can heat much of your domestic water with a well-designed system.

In freezing climates, some form of freeze protection is necessary, either through an antifreeze loop or by draining the collector when temperatures dip below freezing. System design strategies can include PV-driven or AC pumps, sensors, controls, and monitoring.

A solar hot water system is generally somewhat more complicated but much more cost effective than using solar electricity to power a conventional hot water system. It also has a lower upfront cost, with systems running in the $10,000 to $20,000 range. Although they require somewhat more maintenance than solar-electric systems, they're well understood, and more and more conventional heating and cooling companies are adding solar hot water systems to their offerings.

Other Methods for Improving Your Energy Picture

Perhaps none of the alternatives in the previous sections are readily available to you. That doesn't mean you can't do something to improve your energy picture, save money, and improve the environment. Or perhaps you're a go-getter who wants to take on as many renewable energy and energy efficiency options as you can. You have many opportunities, as you find out in this section.

There's no one right way to approach improving your energy picture. Take the time to research your options, take a close look at your situation and resources, and make choices that work for you and your family. What's perfect for one family or house may be impossible for another. Although one may be able to afford one large, expensive system, another may only be able to take on multiple small projects.

Investing in green power

Being realistic about how effective the technologies are at different scales is worthwhile. As with most things, going larger means reducing the cost per unit. A typical 1 megawatt turbine is about 150 feet in diameter and costs more than $1 million. It can supply electricity for about 300 homes, if sited in a good wind resource. If those 300 homeowners all have average residential wind resources, they may spend $80,000 each for systems to generate all their electricity. Three hundred times $80,000 is $24 million. So if you were an energy god, you'd invest in utility scale wind, not residential wind.

This isn't to discourage you from investing in small wind. In the grand scheme of things, society needs small-scale and large-scale renewables. It's not a contest between different sizes or between different technologies. People need renewables at all scales, just as they need everything from skateboards to bullet trains in the transportation world. In all cases, small is likely less cost effective, but as the economist E. F. Schumacher once pointed out, smallness has other beautiful values, such as diversity, variety, interdependence, and local reliability.

If you have a poor wind resource, a thin wallet, restrictive zoning, uncooperative neighbors, or just a lack of interest and ability to have your own wind-electric system, investing in big wind can be an option for you. This investment can take a few different forms:

✔ Many utilities offer *green power* purchase options, often simply a check box on your bill where you can contribute to the green fund. In other cases, you can opt to pay a higher rate and purchase some or all of your home's electricity from wind farms or another renewable resource.

✔ Buying renewable energy credits (RECs), sometimes called *green tags,* is another way to invest in big wind. Various brokers buy and sell these tokens of greenness, which help finance renewable energy installations.

✔ Buy into a community wind project. These are much more common in Europe than North America, but the idea is starting to catch on. This can be a few farmers each putting in a few hundred thousand dollars to buy, install, and operate a wind-farm scale machine. Or it may be a community group raising money for a turbine to offset the local energy usage. Another option that's emerging is private developers looking for investors or partners in a single large wind turbine.

✔ If you have excellent wind on your property, consider developing a small wind farm yourself or leasing land to wind developers.

Cleaning up your transportation scene

Although wind-energy systems make electricity, looking at your total energy picture is useful if your interest is environmental, financial, or both. Transportation accounts for about 27 percent of U.S. energy use. And though transportation is perhaps a smaller segment of your personal energy use (because moving goods about via truck and train is a major energy drain), it's still a big piece of the pie.

First of all, consider how many vehicles you have and how much you drive. You should do a transportation energy assessment, too. Figure out your miles, your gallons, and your maintenance and repair costs. Get a good idea how far you travel and for what. Then before you start dreaming of that new plug-in hybrid, think about how you can be more efficient with your present transportation dollar.

Here are some strategies to reduce your transportation energy use:

✔ Reduce trips by combining errands.

✔ Carpool with others at work, meetings, clubs, and recreational activities. Look for Web sites that facilitate carpooling.

✔ Bicycle and walk more, and you'll have the added benefits of exercising, seeing your town or city in a different way, and maybe getting to know your neighbors better.

✔ Highway speed electric vehicles aren't widely available now, but lower-speed alternatives are:

   • On the low end are pedal-electric bikes. With electric assist, these can extend your bicycling range so that you get out of the car and onto the bike more.

   • Neighborhood electric vehicles (NEVs) are lightweight 35-mph or lower vehicles that could actually handle a large portion of American car trips. Some of these are fully featured and closed in, whereas others are more like simple golf carts.

✔ Try to get rid of surplus fuel-powered vehicles. Think about one multi-purpose vehicle instead of maintaining several more specifically purposed vehicles.

✔ Use mass transit when possible. Though sometimes these are inefficient boondoggle projects, mass-transportation projects do exist, and you only improve them by using them.

✔ Stay home. Explore options for part- or full-time telecommuting. Even working from home one or two days a week makes a difference. If you don't ask your boss, the answer is automatically no.

Simplifying your home and your life

When I speak with private clients about their home energy use and renewable energy, I like to get in before they design the house. At that stage, I can suggest the single most effective way to shrink your financial and environmental footprint: shrink the size of your house design.

Big houses use big energy. Small houses can be wasteful, too, but they use fewer materials, and if you pay attention, they can use much less energy than larger houses. They also have less *embodied energy* — the amount of energy it takes to build and transport the materials. Shrinking your building's footprint shrinks everything about it: cost, embodied energy, ongoing energy costs, maintenance, and headaches — not to mention the time you spend vacuuming and making your house presentable for all those friends you're having over to show off your renewable-energy home.

Riding the wind

Right now, your main options for wind-powered travel are sailboats and hang-gliders. But in the future, I hope to see a direct tie between wind electricity and transportation. Moving the transportation sector toward electric vehicles and increasing the renewable portion of the generating portfolio looks like the best road to transportation energy sustainability. Wind farms and electric vehicles are a practical alternative that people can look to today, and this combo not only reduces energy use but also cleans the air.

To consider simplifying your life, ask yourself questions like these:

- **How big of a house do you need?** Shifting your design from 2,500 square feet to 1,500 square feet (or choosing a smaller home rather than a larger home, if you're buying an existing house) can make a big change in your dollars, energy, and environmental picture.

- **Do your cars actually make your life faster or slower, and which do you want?** As a youth, I once read that if you were to count all the time it takes to earn the money to buy, license, insure, maintain, fuel, and drive cars, you're really going less than 10 miles per hour when you drive. This isn't necessarily intuitive as you race to work at 70 mph so you can pay for the car. Although your job may not give you the flexibility of a shorter workday and a two-hour bicycle commute, I think many people would be happier and healthier if they chose not to drive.

- **Is fast food really fast?** Truly simplifying your diet can save time, money, stress, and illness. Prepackaged and prepared food may seem more convenient, but you buy that convenience with your money, which translates into your time to make the money. Eating a simple, hand-prepared diet may seem time-consuming — but not if you factor in how much time you spend working to pay for restaurant meals and packaged foods.

- **How much labor do labor-saving devices save?** Not only do you have to work to pay for electronics and appliances, but you also have to take time to figure out how they work, take time to use them, and take time to repair, replace, and power them. I'm not suggesting that everyone go back to caveman tools, but it's worth questioning how many things you own and whether the net result is a true savings in your vital energy. So question all energy-using appliances, and see whether you can do a task manually. Try using a shovel instead of a backhoe, sharpening your pencils by hand, and using a wire whisk instead of an electric mixer. In short, seek out simpler ways to do things.

 ✔ **What do you truly need, and what do you merely want?** Life is short, and I doubt that many folks on their deathbeds say, "I wish I'd updated my cellphone" or "I never got to own a convertible." When people look back on their lives, they look at things that are of higher value — friendship, family, society, the planet, as well as personal integrity and accomplishment. Simplifying your life can move you toward all these and away from the superficial things in life that distract you from these core values.

Use it up, wear it out, make it do, or do without. Make the most of the stuff you already have. As I get older, I conclude that stuff is, many times, more burden than benefit.

This book is about wind electricity, but for me, it all comes back to making your life more valuable. Every time you make an energy decision, ask, "Am I fooling myself, or will this truly make my life better?" Check your values and goals, and look at the whole picture before deciding, including all the costs and effects of your decision. You may be surprised to discover that a simpler life is a more energy efficient life, a more sustainable life, and a happier life.

Living on a little

Some of my good friends live in less than 400 square feet of space. Sounds tiny, but their life is actually more full and alive because they're not tied up with maintaining a large home. And because they didn't have to put themselves in debt for 30 years to build their modest home, they have money to live on and have been able to become semiretired at a modest age. Because of their simple lifestyle and a decision to get back to basics, they have only one regular bill — the cellphone bill.

Now you may see these friends as extreme, but they have big lessons to share. Examining your motivations and looking deeply at what actually makes a life higher quality may bring surprising answers. In my experience, people who choose simplicity and people who live simple lives by economic necessity are actually happier and have better lives. I see this regularly in my work in Central America, where folks who may look poor to people from the U.S. actually have very rich lives. Because they're not running about acquiring and paying for all the conveniences that many North Americans strive for, they have more time for family, friends, music, laughter, study, recreation, and following their interests.

Part III
Assembling Your System

The 5th Wave   By Rich Tennant

©RICHTENNANT

NUCLEAR COOLING RODS ACCESSORY KIT

"Hold off on that. I think we're going to get a wind-energy system."

In this part . . .

This part focuses on system design. Chapter 12 covers who will be on your project team. Chapters 13 and 14 look in detail at wind generators and towers. In Chapter 15, I cover the other system components in detail, and Chapter 16 wraps it up by running through the design process.

Chapter 12

Gathering a Team of Experts — or Going It Alone

* *

* *

Successful wind-electric systems use tall towers and large diameter wind-electric generators. Designing and installing such systems requires mechanical and electrical skills, which don't come naturally to most people. Consider the risks and explore your abilities and experience before deciding how to proceed, either with a professional or on your own. I take you through the decision-making process in this chapter.

Considering a Few Issues before You Move toward Wind Electricity

A good wind resource, an energy-efficient home, plenty of space for a tall tower, and permission to build a system are important to have before you take the big leap into wind electricity, but they're not all that's required. To qualify for wind-electric system ownership, you absolutely need either a hands-on mindset or a willingness to open your wallet upfront and on a regular basis. In this section, I give you some final issues to consider before you officially start the process of installing a wind-electric system — whether you choose to hire someone for the job or do it yourself.

The difficulty of installation

Installing a wind-electric system is perhaps comparable to installing a complex modern heating system. Well, it's actually harder, because I've never seen a heating system installed on the top of a 100-foot tower. This isn't to say that you can't develop the skills and do the work. But you need to be honest about your abilities, time, and commitment to the project.

Most people can choose and buy a toaster. You know what the goal is — golden-brown bread toasted to perfection and ready for that orange marmalade. You may want the chrome finish, perhaps with classic 1950s lines or the modern shapes and styles. You may want to avoid *phantom loads* (electrical loads that continue to use energy when they're switched off; see Chapter 7), and perhaps you want a bell, buzzer, or other indicators to let you know when the job is done. Buying it is the hard part. Installation amounts to plugging it into the outlet next to your microwave.

Installing a dishwasher is a big step up from installing a toaster. You need to buy the right model to fit the space in your kitchen cabinets. You need to deal with physical connection to the cabinetry, leveling, and vibration and noise issues. You need to make electrical connections (I hope through a switch, so you can turn off the phantom load). And you need to deal with the plumbing, which is sometimes not so simple. Many people wouldn't attempt this task, leaving it up to appliance and building professionals to make sure the dishwasher is safely and effectively installed.

Designing and installing a wind-electric system is much, much more difficult than the dishwasher job. Don't even consider doing it yourself if the dishwasher job is out of your realm. If you'd hire out the dishwasher installation, you definitely want to hire out the wind-electric installation.

In my years of teaching contractors, homeowners, and others about wind electricity, I've realized that although I can teach the key design concepts in the brief workshops I do, I can't teach all the skills necessary to be a successful wind-electric system installer. You have to pick up most of these skills on the job, and they include the skills of designer, contractor, tower jockey, and electrician. If you feel you may be a good candidate to work on your wind-electric system, see the later sections "Preparing to Do the Installation Yourself" and "A Little Help: Taking the Middle Way."

The presence of serious hazards

Wind-electric systems have a full complement of hazards:

- ✔ Tall towers with unforgiving gravity
- ✔ Electricity, with shock and fire potential

  ✔ Perhaps batteries — heavy, poisonous, and powerful

  ✔ Rotating equipment that can break and fly apart

Wind electricity is not something to mess with if you're not a serious, hands-on person ready to do the homework (especially on safety) and get your hands dirty — or someone with deep enough pockets to hire the needed help. Are you up for the job and the risk? I discuss safety in Chapter 17.

Maintenance requirements

There's no such thing as a maintenance-free wind-electric system. The equipment is spinning, and all spinning equipment requires maintenance. Your turbine is in a severe environment, which means it encounters dirt, crud, wear, and tear. You'll have trouble sometimes even if you do the regular maintenance, but you almost guarantee problems if you ignore the service requirements. You should plan on at least an annual checkup; very windy areas or low-budget wind turbines may need a check twice a year.

Are you the kind of person who takes care of things or the kind of person who just wants them to work magically? Wind electricity is magical in its way — watching a natural force provide electricity for you is remarkable. But if you expect it to be free, cheap, and easy, you'll be disappointed. (See Chapter 10 for more information on calculating the potential costs of maintaining a wind system and Chapter 19 for the maintenance tasks required.)

The probability of trouble, even with regular maintenance

Wind-electric systems on the utility scale are mature technology with high reliability. But even the wind farm turbines require regular maintenance and have regular challenges. A wind farm of any size has a crew that rotates continually from machine to machine making sure everything is working properly.

With only one machine, you have less responsibility and potential for trouble than a utility-scale wind farm does. But home-scale machines generally aren't very mature technology, so users often end up being the beta testers for the manufacturers. Many manufacturers are trying to increase performance, experimenting with ways to increase power and energy. But the best manufacturers instead focus on improving reliability — it's the biggest factor in achieving the goal of generating kilowatt-hours for the long term.

In addition, most American consumers want products to be both good and inexpensive, which is a hard combination to produce. The problem is that many buyers lean more heavily on the *inexpensive* part of the equation.

The end result is that home-scale wind-electric systems are rarely without issues — you see failures, erratic problems, and hard-to-identify conditions. Finding a system that's issue-free is hard — unless it isn't installed and operating yet (they all look pretty good in the box!). If you're not ready to deal with such issues, you're not ready to own a wind-electric system.

Consider the vintage of the wind generator you're thinking of buying. Do you really want to buy the first car of a new model? Perhaps waiting until the folks on the design team and on the assembly line have worked out the bugs is a better idea. If you choose a largely untested, cutting-edge system, you need to be prepared for some entertaining failures and frustrations. Most people should wait for a track record on a specific machine.

Deciding Whether to Have Professionals Install Your System

In the early days of wind-electric systems, most if not all systems were installed by the homeowner. These folks came from all walks of life, but they were all hands-on. The industry wasn't developed to the point where you could hire Joe or Jane Wind Installer to do the job for you, and many folks couldn't have afforded to pay for such service if it had been available.

Today the situation is very different. Most home-scale wind systems are installed by professional installation companies, and that will become more and more common in coming years. There are rumors of major mechanical equipment chains in North America getting involved in wind-electric systems, with branches hiring local construction crews to do their installations, so installing these systems will become standard practice for specialty construction firms in the future.

This isn't to say you shouldn't install your own system, *if* you have the skills and situation to do it. But those skills make you an increasingly rare beast, so look carefully at the advantages of an installed system. I describe them in this section, along with a couple of disadvantages.

The advantages

Installing a wind-electric system involves many different tasks, specialized equipment and tools, and a broad range of knowledge and skills. The advantages of hiring this installation out are many, as you find out here.

A turnkey setup

If you're careful when choosing your contractor, you should end up with a *turnkey system* — in other words, you just "turn the key" to start it, and you don't have to construct or commission it. Frequently, owner-installers end up with a lot of head scratching, troubleshooting, and fine-tuning of a new wind-electric system. If you hire it out and get a firm price, all this trouble and expense is on the installer's back, not yours.

Be prepared for the process to be complicated, even with an experienced installer. The industry is small and immature, and glitches occur regularly even with professional installation, so you still may need considerable patience. Don't expect a professional wind-system installation to be like someone blacktopping your driveway — in and out in a day — because you may be unpleasantly surprised. But if you hire well and give adequate time, you can get a great system with few installation headaches.

Equipment, installation, and operation warranties

If you hire out the installation of a system, you should end up with a system installation and operation warranty from the installer in addition to the manufacturer's warranties for the individual component products. Be persistent about getting this installation and operation warranty! It means the installer agrees to stand behind his or her installation work. It also generally means that the installer handles any warranty issues with the specific equipment.

Some states that offer incentive programs for wind power systems require installers to offer a five-year warranty on the system as a qualification for incentive funds. If your state doesn't offer such a program, a five-year warranty is probably too much to expect. But you can reasonably demand a one- or two-year warranty on the installation.

Don't underestimate the value of the installation and operation warranty! It may be worth thousands of dollars in an industry full of small companies with relatively new products. Because the industry and equipment is constantly changing, bugs frequently need to be worked out with products and systems, and having a professional lined up to handle them saves you lots of headaches.

Service and support

Another benefit of hiring out your wind-electric system installation is that you have a professional who is familiar with your system's service and support needs. This benefit will outlive the warranty with a good installer, and even though you have to pay for the help, having a professional connection is a huge step above starting from scratch trying to find and qualify an experienced technician.

Ask installers whether they offer a service contract, an extended warranty, or any other prepaid means of covering the necessary service.

A relative lack of stress

From the first design meeting to the ongoing maintenance, hiring the job out means you're essentially shifting stress and responsibility from yourself to your wallet. You may be part of choosing the big components but not the small. Typically, a contractor presents you with one to three options of major components, but more often this is a size choice.

When you have an experienced contractor on the job, it's the contractor's job to shoulder the responsibility of sourcing and installing good equipment and making sure it's productive for the long haul.

The disadvantages

You may think hiring out your installation is all benefit, but like most things, it also has downsides. The two primary disadvantages of hiring out your wind-electric system design and installation are cost and lack of hands-on knowledge about your system, which I discuss in this section.

Cost

Hiring people to do work for you costs money. In the case of a wind-electric installation, you can expect the labor portion to be 20 to 30 percent of the total cost of the system. I encourage you not to get hung up on this figure, because you still have to bear that cost yourself if you do your own work, either through hiring consultants or spending your own time doing the brain-work, legwork, and grunt work. That's not free, so look at how much you can earn in your own line of work and ask yourself how much you gain by taking time off to do your own installation.

Don't underestimate the value of an experienced contractor's brain. These folks spend most of their working time designing and installing systems, so it comes naturally to them. They know the tricks of the trade, which saves you from learning them through your own mistakes.

Get a firm price upfront so you know just what the job costs, and ask for at least a basic breakdown of equipment and labor costs. Then make a wise decision based on your own abilities, time, and earning power.

Less personal knowledge

If you hire out your system and let a professional do the whole thing alone, you lose out on the benefit of finding out more about your system. Sure, you can read the manual and ask questions of the contractor. But nothing beats hands-on involvement to give you direct knowledge of the system you have to live with.

Even if you hire a professional to source the equipment for, design, and install your wind-electric system, you should maintain some involvement in the project. This arrangement gives you the best of both worlds — the benefits of self-installation without the liabilities. You avoid full responsibility for (and major headaches associated with) the design and installation, and you also get out of handling warranty claims. In return, you get

- ✔ The knowledge and abilities of the installer

- ✔ The experience of getting your hands and brain involved

- ✔ Warranty and service coverage

- ✔ Someone you can call with questions or concerns

I see this lack of knowledge as a major disadvantage of buying an installed system and hope you consider the middle ground of helping your contractor (which I describe later in this chapter). You may actually pay the same or more to "help" because having homeowners involved doesn't necessarily save the contractor time or money. But having a more intimate knowledge of how your system works is money well spent.

Finding and Hiring an Installation Pro

Unless you're extremely confident in your installation skills, hiring a pro is likely your best option. This book can't make you a wind-electric system installer, but it can give you the information to choose a good pro and avoid the bad ones. If you choose carefully, you can get a high-quality installation that will give you reliable electricity for years.

When hiring someone to install your wind-electric system, the whole goal is reliability — any other claim or focus from the marketing guys is irrelevant. This means that you're not looking for a cheap installer; you're looking for competence. You're looking for conscientiousness. And you're looking for someone who will be there when you hit the challenges. Your installer will probably also be the one you call on to do routine maintenance on your system and the one you call when you have questions, glitches, or serious problems (see Chapter 19 for more about maintenance).

In this section, I explain the process of finding a professional installer and describe some traits your ideal installer should have.

The process of looking for a pro

Finding your wind-electric installer is a process, and it may not be a short one. The end goal is to sign a contract that specifies what, how, and when you'll start generating wind electricity — and at what cost. If you want to reach this goal, you need to find someone who has been down that road before and has left a trail of satisfied wind-energy customers.

Installers go by a number of titles — common names include *wind-energy contractors, renewable energy contractors,* and even *solar contractors.* You can find these pros through the Yellow Pages, renewable energy associations, and your best source, other wind-energy users. Some manufacturers list available installers on their Web sites, but quality varies. You may find brand new installers who've never installed a turbine alongside very experienced installers with dozens of projects under their belts.

Ideally, you want a single party to do both design and installation. With a larger installation, you may have a separate engineering company do the design, but then the designers normally contract with another company for installation.

The traits you should seek in a pro

When you start looking for a pro installer for your job, you should look at three basic things: experience, cost estimates, and working style. I discuss each trait in this section.

Searching for a pro with experience

Experience is almost everything with wind-energy system installers. Because the industry is immature, you don't see many certifications or training programs. Most training happens on the job, and one gauge you can use is how many jobs the person you're talking to has been on.

At a minimum, find someone with several installations completed and get the names and contact information for their clients. Call the owners and ask for a frank assessment of the contractor's abilities. And ask how they feel now about their investment in wind electricity. Don't skip this step. Not only are you putting out a big chunk of change, but you're also trusting your hopes, your home, and a bit of your future to the installer. Make sure you'll get what you expect from him or her.

 Look for installers who live with wind-electric systems themselves (though finding someone like this isn't always possible). These folks know the pitfalls, struggles, and benefits. You won't likely hear them hyping wind electricity, because they have to live with the reality.

Finding a pro who's honest about the costs

Dreamy thinking about costs can set you up for major disappointment. Dealers I know and trust are upfront about the costs involved. When I speak to a client, I like to tell them that a home-scale wind electric system will cost between $20,000 and $200,000 and that $20,000 will be tough to attain in most cases. It's much more typical to spend $30,000 to $60,000-plus.

If you're not ready to spend at least $20,000 on a wind-electric system, my advice is to just let wind power go for now. Cutting corners and going halfway make sense in some places in life, but this isn't one of them. Your goal is probably to save money or the environment, and you'll be ineffective at both if you don't spend enough to make a reliable, long-term, and productive system. See Chapter 11 for other options that may help you attain your goals.

You may think that a pro who trumpets the high cost is out for your wallet. Well, these folks need to make a living, of course, but I have the greatest respect for installers who are very clear upfront about the level of investment required to make these systems work. (Flip to Chapter 10 for details on calculating the various costs. There, I show you how to calculate the cost of a wind-electric system per kilowatt-hour. Using that basis, you'll likely find that companies that are pushing "low-cost" systems are actually selling the most expensive electricity out there.) And in fact, larger systems will be much more cost effective than smaller ones, in general. I often talk to clients who wish they had invested more in a larger system on a taller tower. Besides the initial investment, what's the downside to having more clean energy?

Making sure you like a pro's style

So you want a pro with experience and a pro you can trust. You also want someone you can feel good about personally. When you own a wind-electric system, you'll have more contact with your installer than you would with your toaster salesperson or your dishwasher technician.

This person and crew will be on your property for several days for design and installation of the system. You'll be on the phone from time to time over the years of its operation, and these folks will visit once or twice a year to do service.

Don't choose someone you can't get along with. Wind-electric systems are hard enough to deal with without personal dynamics getting in the way. Spend enough time with the contractor, his or her employees and clients to get a feel for who they are and how they operate. If you're a tree-hugger, find someone who shares your ideals. If you're a math geek, find an engineer, someone who speaks your lingo. In many cases, these people will not only become part of your renewable energy support network; they may become your friends as you move toward realizing your dreams, and they help along that path.

Preparing to Do the Installation Yourself

Wind-electric system design and installation is not a Saturday afternoon hobby project. But if you're determined and willing to spend the time and money, you can do the installation yourself. You'll need to go well beyond this book to gain the skills and experience necessary, as you find out in this section, but it's certainly possible.

In some cases, doing your own installation can affect whether you're allowed to connect to the utility grid, or it may limit access to incentives. Both questions are very specific to the state and utility, so make sure you know the rules and interconnection standards before you decide to do the installation yourself.

Getting training and finding some partners in crime

When you start a project, you need to assess not only your budget and plans but also what you bring to the scene. Are you an electrician or an accountant? A machinist or a retail salesperson? What are your hobbies? What are your abilities? How quickly do you pick up new skills?

Watching hundreds of students come through the programs I teach, I see some very promising people with lots of construction, electrical, and mechanical experience. Finding someone who has all the pieces and can seamlessly transition into doing a great job at designing and installing these systems isn't very common. This is undoubtedly even truer with the average homeowner, so be honest with yourself and figure out which skills you have and which skills you don't. Take advantage of the former and upgrade the latter if you're serious about this job.

Training for wind-electric system design and installation is not easy to come by. You won't find it at your typical community college or technical college (though you may find classes on basic electricity and wiring, which can be helpful background information). You have to search out specialty programs at generally small educational organizations such as Solar Energy International (www.solarenergy.org), the Midwest Renewable Energy Association (www.the-mrea.org), or Cape & Islands Self-Reliance (www.reliance.org).

You'll be hard-pressed to find a full college program on home-scale wind-electric systems — most organizations only offer a few weeks of wind electricity training. So you'll need to broaden your search and get training or experience with the following:

- ✔ Electrical work
- ✔ Mechanical work
- ✔ Design concepts

Preferably, this experience involves actual installations. It's probably too much to hope that you can install a system that matches the one you plan to buy, but try to find the opportunity. Though wind-generator manufacturer training is usually for professionals, you may be able to get involved. Check manufacturer Web sites, or just give manufacturers a call and ask when the next training is scheduled.

Local wind-energy users also may be allies, especially if they installed their own systems. These people are usually advocates or activists — some are maniacs — for wind energy. Many will jump at the opportunity to volunteer on your job, sharing their experience, abilities, and enthusiasm. Take the time to visit their sites and see what they've done. That can help you decide how much to trust them, and you pick up info from seeing their systems and interviewing their families.

If you're new to wind energy, buy your equipment from a supplier who can supply more than just equipment. Find a small company that provides excellent support and ask upfront whether they'll walk you through the design and installation process. Ask whether you can call them when you have questions or are confused about how to install something. *Home Power* magazine periodically publishes an article surveying serious machines from established companies; tracking down the dealers for these machines can yield good results, though you still need to carefully research the dealers you find. (Chapter 16 has more information on finding qualified suppliers.)

Having a savvy supplier on your team can make a huge difference. This may be an Internet-based company in another state with great customer service. Or it may be a local installing dealer. Even if you don't hire the company to do the installation, the supplier can be a great help. Don't expect this to be free, though. Your gear prices won't be rock-bottom from these people. And they may even charge a design or consultation fee. Consider the costs and benefits (my money's on having experienced people to turn to when you don't know what to do).

Don't try this at home? Heeding a few warnings

Do-it-yourselfers are intrepid, innovative, and determined. These are great qualities. But there are limits to what you can wisely do with a low level of experience. In my book, the following are no-nos:

- Designing a wind-electric system without study, volunteer experience, and mentors

- Installing a tilt-up tower without previous training

- Climbing towers without the proper gear, instruction, and experience with a veteran climber

- Messing with electricity as if it's plumbing (the risk of amateur plumbing is somewhere between some drips and a minor flood, whereas the risk of amateur electrical work is shock, electrocution, and fire)

Am I saying that you shouldn't design and install your own system? No. Hundreds of people have done this, including me. Perhaps my biggest quali- fication was stubbornness, but time, study, persistence, and several failures have taught me the lessons. Perhaps this book will save you from the biggest mistakes, and you can learn from others' and mine. But you need much more preparation than just reading a book. Think in terms of how much heating- system contractors know. They didn't start their business after reading a short book. They have weeks or months of study and years of on-the-job training.

A Little Help: Taking the Middle Way

So perhaps you've realized that this job is not as easy as you'd hoped. Getting real is painful, but it can lead to better things. You may decide to get the training and experience necessary to do this job yourself. Or you may decide to hire a professional to do the whole job.

But you do have options between these extremes. You can improvise from the suggestions in this section, depending on your circumstances. In general, I think that more heads are better than fewer. From time to time, I've worked on a crew of six or seven that installs wind generators. Several of us have enough experience to lead the jobs, but we choose to work together because it's fun and instructive. There's no downside to having more experience and expertise than you need, as long as you keep the egos in check.

Partnering with your supplier as mentor

In the earlier section "Getting training and finding some partners in crime," I mention using your supplier as a source of information as you do your own system. You can take this further and involve your supplier in the design and installation process directly. On one end of this option, you can ask a professional installer whether you can help him or her do the job. Of course, you should be aware of how this impacts your supplier and the budget.

You may have seen the humorous sign hanging in auto shops — "Hourly rate: $55; $65 if you help." There's a bit of truth in this. Make sure you'll actually be helping or that you're happy to pay the extra price to be involved. Ask your supplier which parts of the process you can be most helpful with and where your involvement may actually increase the cost. For example, digging ditches is an obvious example of helpful involvement, but intricate wiring may be an area where the supplier spends more time explaining things to you than getting the work done. Larger contracting companies are less likely to accept your involvement, whereas mom-and-pop shops tend to be more open to it. Most should be open to your doing preparation or cleanup before or after they come on site.

Before you put on a hard hat or start fitting yourself for a safety harness, ask how your participation in installing the system will affect warrantees or insurance issues.

Thinking teamwork

If you don't have the experience and skills to tackle the job of installing a wind-electric system, don't overlook your circle of acquaintances. Many people are excited about wind energy, and you may find that assembling a volunteer crew that shares your enthusiasm isn't hard. Expertise is hiding all over the place. For example

- ✓ Anyone with strong handyman skills can be of great help. That farmer down the road whom you've known for three decades knows how to fix his tractor and keep his baler running, and he built and maintains all the buildings on the farm. He may be a great addition to your team.

- ✓ Having someone with electrical knowledge, experience, and credentials on your job is essential. And you may luck out and find someone who wants to know more about these systems and is willing to help you at low or no cost. If not, hire an electrician to be part of critical stages in design and installation.

✔ Friends can be a great help or a great burden. Find out what their abilities are and take a look at their personalities before signing them up for your project. You need sharp, cautious, humble people, not loud, aggressive, overconfident, and overbearing people. Find people you enjoy working with and people willing to be self-critical and aware. Wind-electric systems suffer when strong egos get involved. Let cooler heads prevail.

Getting a professional review at crucial stages

If you decide to do some or all of the work yourself, I strongly urge you to get a professional review at key stages in your project. Your supplier may offer this function for a fee, and this can work well if you've chosen a high-quality supplier. You may also want to consider an outside consultant so you can get another perspective. I recommend bringing in this expertise in at least two stages of the project:

✔ After the design is complete but before you purchase the gear

✔ After the project is complete but before it's commissioned

Preferably, you'll add a couple more stages:

✔ At the initial design stage

✔ At the beginning and in the middle of construction

Using available expertise is a wise choice. Don't get stuck on the cost of this service. Think about the potential cost of not using such a service. If you make a design decision that results in an underperforming or dangerous system, you'll have saved money upfront but lost much more down the line. Instead, invest well upfront for long-term safety and performance.

Chapter 13

Weighing Wind Generator Options

*T*he wind generator, or turbine, is the main attraction in a wind-electric system. As you find out in Chapter 3, it can't do anything on its own. But it's what captures the wind energy and uses it to make electricity. And it's an important choice when designing your wind-electric system, as you discover in this chapter. Here's why:

✔ You need to be clear about the general scale of wind turbine useful for your project. Is your energy appetite sailboat-sized, home-sized, ranch-sized, or utility-sized?

✔ You have to look at the various configurations available in your size class. As with anything, educating yourself about the options helps you make a wise choice.

✔ You have to make a real-world choice, which includes evaluating your site and wind resource, your own abilities and goals, and your budget constraints or preferences.

If you take all these factors into account, you'll be set to buy a wind generator. Choosing carefully can mean the difference between a life of pain and a life of gain.

Surveying Wind Generator Sizes

Wind generators come in different sizes. The size of the machine you need is determined by how much energy it can produce, which is directly related to the area that the blades sweep (called the *swept area*) as they turn. This area is the *collector,* and if you want more energy, you need more collector area. Essentially, *small swept area* means small energy production; *large swept area* means large energy production. (Head to Chapter 5 for more details on swept area.)

The size of a generator's swept area is the biggest factor in determining how many kilowatt-hours you get, and in the end, kilowatt-hours is the measurement you should use to determine how large a wind generator you need. For example, if you want 1,000 kilowatt-hours per month and you have a 9 mph average wind speed (see Chapter 5 to understand what this means), you'll look for a wind generator that will deliver that much energy in your average wind speed.

Note that it's very common to specify wind generators by their peak or rated power in watts or kilowatts. (Chapter 4 explains the difference between watts and watt-hours, and Chapter 5 discusses rated power.) This is a big mistake that leads to much confusion. Because power varies with wind speed from moment to moment, rated power is not a good indicator of energy, so it doesn't help people understand just how much energy they'll get.

You can find out how to predict how many kilowatt-hours a wind generator makes in your average wind speed by flipping to Chapter 5. But if you can't determine the number of kilowatt-hours, you'd do best to identify wind generators by swept area. The *swept area* is the circle that the blades describe, and that square footage has a direct relationship with how much energy your wind generator can generate. (Rotor diameter is handy shorthand, but it doesn't give you a clear view because the swept area quadruples if you double the rotor's diameter.)

Four basic sizes of wind generators based on swept area are worth identifying (see Figure 13-1 for a comparison in scale). These sizes are sailboat-scale, home-scale, ranch-scale, and utility-scale. Being clear upfront about what you need and want saves you, your dealer, and anyone else involved time and trouble.

Sailboat-scale turbines

I call wind generators in the 2- to 7-foot diameter range *sailboat turbines* — their swept area is about 3 to 40 square feet. You can also call them *micro turbines*. This scale of wind generator gives you from 50 to 1,000 kWh per year in a 10-mph-average wind resource, or 0.15 to 3 kWh per day.

The typically wasteful American home without electric heat uses 25 to 30 kWh per day; because these wind generators provide at most only about 10 percent of that amount (less than 5 percent if the home uses electric heat), they don't provide significant energy for normal homes. These turbines are truly micro — tiny contributors of energy. So you can see why that this scale of turbine is more useful for boats (hence the name), parked RVs, very efficient cabins, or other small loads.

The installed system cost for this range of turbines is perhaps $3,000 to $30,000. Of course, this price depends heavily on the type of tower (see Chapter 14) and the system configuration (see Chapter 9). And how much energy these generators produce depends on a crucial factor — you guessed it, the wind resource.

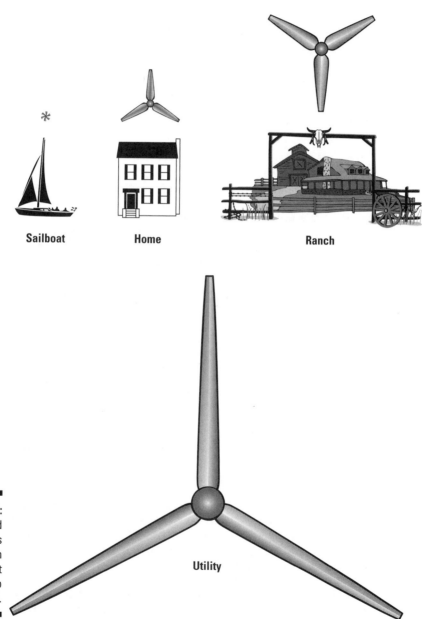

Sailboat

Home

Ranch

Utility

Figure 13-1:
Wind generators range from sailboat size to utility size.

If you have a tiny load, a sailboat-scale turbine may be perfect; just be realistic about how much energy you need and how much swept area and wind resource you need to make it.

Home- and ranch-scale turbines

I expect that if you're serious about wind energy, you want it to make a serious impact on your energy usage, utility bill, and environmental footprint. If you're using 30 kWh per day, you need a home- or ranch-scale turbine, one that can provide a significant portion of your electricity with your wind resource. I explain the differences between home- and ranch-scale turbines in this section.

Home-scale turbines

Wind generators in the *home-scale* class are between 8 and perhaps 25 feet in diameter (the swept area is about 50 to 500 square feet). They provide between 1,000 and perhaps 15,000 kWh per year on a 10-mph-average wind resource, or 3 to 40 kWh per day. The top end of this range is a serious amount of juice. It's actually four or five times as much as I use every day in my energy-efficient, off-grid home, so I'm salivating just writing these big numbers.

You won't be popping one of these wind generators on your boat, and installing one isn't cheap or easy. You need a large, tall tower (see Chapter 14) sized to be able to support a turbine with this swept area and get it up into the good wind.

Systems in this size category cost in the $30,000 to $100,000 range installed, depending on turbine and tower size. Your mileage (well, electricity output) will vary with the quantity of wind available (in other words, the wind resource) and the quality and configuration of the system.

Ranch-scale turbines

Ranches are big spreads, and not surprisingly, they use large amounts of energy. I bet you've figured it out by now: If you want large amounts of energy, you need large energy collectors.

Turbines in the ranch-scale class run from about 25 to 70 feet in diameter (the swept area is about 500 to 4,000 square feet). In a 10-mph-average wind resource, they produce between 15,000 and 75,000 kWh per year, or 40 to 200 kWh per day. The top end can run the ranch house and barns and pump water for a big herd of cattle.

A top-end ranch-scale turbine can run a substantial village in the developing world, or a small group of super-efficient homes in a co-housing community in North America.

Installed cost for these turbines ranges from about $60,000 to perhaps $300,000, again depending on size, tower height, tower style, and system configuration. As with all wind energy systems, production depends on the wind resource that the turbine sees at hub height (the *hub,* the center of the blades, is a common point to measure tower height from).

A category that usually falls between ranch scale and utility scale is what's often called *community wind.* Turbines may be in the 60- to 150-foot diameter range (that's about 2,800 to 17,700 square feet of swept area), and they cost perhaps half a million dollars and up. These machines may be installed by groups of citizens, nonprofit organizations, or even governments, and they may make enough electricity for 100 to 200 homes. They provide energy to the various entities at a better cost than home-scale machines on each individual house and can relieve the owners of the individual responsibility. You see this kind of setup more often in Europe, but I'm glad to see it cropping up here and there in the United States as well.

Utility-scale turbines

Utility-scale turbines are well beyond the scope of this book, but they're worth mentioning because people interested in small-scale wind often end up interested in large-scale wind as well. My work is with home-scale machines, but I'm a strong supporter of wind farms as well because I believe we have to change the way we make electricity in this world if we want to leave it in decent shape for our grandchildren.

Utility-scale machines run from about 150 to 400 feet in diameter (about 17,700 to 126,000 square feet of swept area), cost upwards of a million dollars each, and power 300-plus homes. If you have that sort of change falling out of your pockets, I hope you'll make a wise investment in the technology and support this worthy industry.

Considering Differences in Wind Generator Configuration

Understanding some basic configuration differences can help you choose a wind generator. Some of these differences are important, but others aren't that relevant. Knowing all the differences can improve your ability to shop.

This section looks at some of the common options you can choose from. In the end, the most important idea is to determine whether a given wind generator will be productive and — most importantly — reliable on your site. (For an introduction to basic wind generator parts, check out Chapter 3.)

I hope you find trusted cohorts to assist you with these decisions (see Chapter 12), because becoming an expert in all the components and technologies in a wind-electric system is a big, big job. Lean on your contacts, suppliers, installers, and others as you go through these decision-making processes.

Often, an important factor in your turbine decision has nothing to do with configuration and everything to do with availability. If you live in North America, you may be able to get a most excellent turbine from Australia, but if you can't find parts, service, warranty coverage, information, and support locally, you're better off with a less excellent turbine made closer to home.

Battery-charging versus batteryless grid-tied

You can ignore many of the distinctions in later sections (assuming you're buying from a reputable dealer), but this one is perhaps the most pertinent distinction between classes of turbines. If you choose to have batteries in your system (either because you're not hooked up to the utility grid or because you want utility outage protection), you need to buy a battery-based system. If you instead choose the more efficient and simpler batteryless grid-tied option, you need to buy that system.

Chapter 9 discusses these options in far more detail, but for the purposes of this section, ask yourself, "What's my goal?" For most people, the primary goal is cleaner, cheaper electricity. But specifically, some people want off-grid electricity or battery backup while they stay on-grid, which means storing electricity in batteries (and not your standard AAs but a large, expensive bank of deep-cycle batteries). Other people want to connect to the grid but have no need for utility outage protection, which leads them to a batteryless grid-tied system.

If you're on-grid but want outage protection when the local utility fails, you need batteries, and this setup is quite possible — just ask for it. But most on-grid systems installed these days don't have batteries and don't have backup capability, because most people live where the utility grid is reliable.

So when it comes time to buy, tell your supplier which you want — a battery-charging machine or a batteryless grid-tied system. These are two entirely distinct system types, and you need to make sure you get what you want.

Direct drive versus gear driven

Most home-scale wind generators today are *direct drive,* which means that the blades are directly attached to the rotating portion of the alternator. When the wind blows, it turns the blades, which turn the alternator and make electricity. *Direct* is a good description of this process — no middleman parts are taking a cut, so you get the most energy out of the wind in the simplest way.

As wind generator blades get longer, you start to get into a situation where using gears to increase the speed to the alternator may be more appealing than building a larger-diameter alternator. At this point, you're looking at a *gear-driven* system. If the rotational speed of the blade rotor is fairly slow, turbine designers may opt to use a gearbox to increase the speed for the alternator. That generally means you need a gearbox. See Figure 13-2 for a comparison of direct-drive and gear-driven designs.

Beware the following disadvantages of gear-driven generators:

✔ As soon as you introduce gears, belts, pulleys, or other ways to increase or decrease speed, you introduce energy losses. Any time you convert energy from one form to another, you lose some of that energy to the conversion process. Some of the energy you're trying to take from the wind goes into heat in the gearbox, and you get less power running down the wires to your house.

✔ You make a machine more complex by adding gears. That means more parts to buy, maintain, and replace. Gear-driven machines have higher maintenance requirements, including inspection, oil changing, and replacement of parts.

Direct-drive versus gear-driven is a point that would affect my buying decision. I definitely recommend first looking at direct-drive machines for their simplicity, reliability, and lower maintenance requirements. If other factors push you toward a gear-driven machine, plan ahead for regular maintenance.

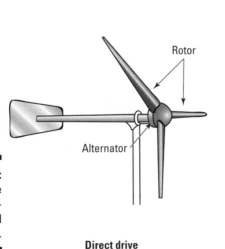

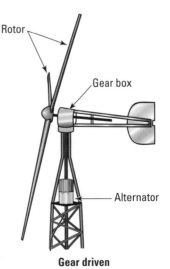

Figure 13-2:
Direct-drive
and gear-
driven wind
generators.

Direct drive

Gear driven

Gear-driven machines may have better electrical efficiency in high winds because of the high speed of the alternator, but they may have lower mechanical efficiency in low winds because of the losses in the gearbox (friction) that are relatively high compared to the power available in low winds. Standalone systems need to prioritize low wind speeds to reduce reliance on battery storage, so direct-drive is more popular than gear-driven in standalone systems.

Horizontal axis versus vertical axis

Prepare for some strong opinions when you start looking into the difference between horizontal- and vertical-axis wind turbines (see Figure 13-3). The *axis* of a wind generator refers to the line in which the rotor shaft rotates.

- A *horizontal-axis wind turbine* (HAWT) has blades that spin on a horizontal shaft. It's like the normal machines you see in wind farms, on homesteads and boats, and in pictures of windmills in Holland.

- A *vertical-axis wind turbine* (VAWT) has a vertical shaft that the blades rotate around. You may have seen egg beater-style wind turbines in experimental wind farms, and you've likely seen the classic homemade ones created by cutting a 55-gallon drum vertically and offsetting it on a vertical shaft to capture the wind.

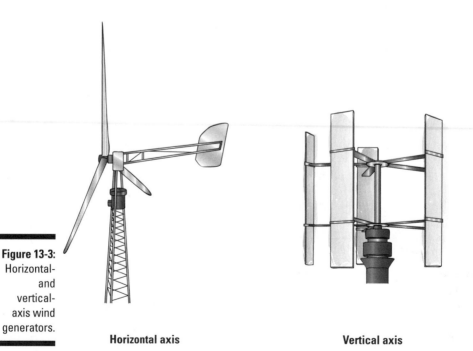

Figure 13-3: Horizontal- and vertical-axis wind generators.

Horizontal axis   **Vertical axis**

In my opinion, this decision is easy: Stick to HAWTs. The market gives a clear indication of what works well, efficiently, and reliably, and almost all machines on the market for any length of time are HAWTs. This tendency isn't because of some conspiracy against VAWTs or because all the companies producing HAWTs are full of idiots who don't know the "superior" value of VAWTs. It's because HAWTs work better and not only produce energy better but also produce profits for the manufacturers because they're reliable, productive machines.

Although producing a good VAWT isn't out of the question, VAWTs will never be as productive for their swept area as a good HAWT. And they can have significant problems with support, bearings, fatigue, startup, and speed control. People seem to think this configuration has major advantages, but these advantages largely turn out to be misunderstandings about wind energy and wind generators. Here are a few VAWT myths:

- ✔ **Takes wind from all directions:** So do HAWTs — no advantage.

- ✔ **Can go on your roof or a short tower:** All wind turbines need a good wind resource, which you don't find in these locations.

- ✔ **Does better in very low-speed winds:** This statement is irrelevant (not to mention untrue) because low-speed winds provide so little energy.

- ✔ **Handles turbulence better:** Turbulence is the enemy, so putting any turbine where turbulence exists is a bad idea. And as my colleague Hugh Piggott quips, "They may be better at handling turbulence because they live in a world of turbulence that they create themselves." No advantage.

- ✔ **Smaller rotor makes more energy:** This claim is nonsensical, as you discover in Chapter 5. Smaller rotors make less energy — the rotor is the collector, and you need larger ones to make more energy.

A prominent feature on the Web sites of most VAWTs today is the "Investment Opportunities" button. These companies are looking for your money, but don't expect to get it — or a lot of energy — back. I wish someone would produce a productive, reliable VAWT, because the technology can work. But the field is so full of hypesters and scam artists that it's hard to see much substance. My recommendation is to steer clear of VAWTs, both for power and as an investment. Even if and when a serious company makes a serious machine, I doubt it'll be better than a HAWT.

Two blades versus three blades

The number of blades on a generator is a significant issue. How many blades should be on your dream wind turbine? With modern, home-scale machines, you generally have a decision between two and three blades (see Figure 13-4).

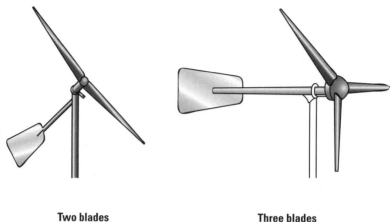

Figure 13-4:
Two-bladed
and three-
bladed wind
generators.

Two blades   **Three blades**

Two-bladed wind generators may cost a bit less, because you pay for only two blades. Unfortunately, they come with a major drawback, which is that they lead to what's called *blade chatter,* a vibration that puts wear and tear on the whole machine (see the sidebar "Understanding blade chatter" for details).

Three-bladed machines don't have this issue, because with three blades positioned 120 degrees apart, you never get a direct imbalance that two-blade systems are susceptible to — which tells you why most machines on the market are three-bladed: They run more smoothly because they're better balanced. They also tend to be quieter because they run a little more slowly. Overall, three blades is a good compromise between optimum alternator speed and sound level. However, there are good quality two-bladed machines on the market that perform well and have overcome the drawbacks.

Scoping out homebuilt wind generators

Perhaps you think that instead of buying a manufactured wind generator, you can build your own. This option is viable if you're a hands-on person who knows how to use tools. The world is full of homebrew wind-generator designs, but most of them are pretty worthless. If you decide to go this route, be careful to find working real-world examples of any design before you put time and money into it.

One prominent example is the work of Scottish wind enthusiast Hugh Piggott. Hugh started out building machines to power his off-grid home

on a remote peninsula in Northwest Scotland in the 1970s. Today, he sells wind generator plans and teaches several courses a year on how to build his axial-flux design. The resulting machines are simple, durable, and productive. You can find more information at www. scoraigwind.com.

Stateside, the great crew at www.other power.com has taken up Hugh's work and innovated with it. They now sell plans, books, parts, and complete machine kits.

The disadvantage to a three-bladed rotor is primarily cost: You pay for three blades instead of two. These machines may suffer some efficiency loss — adding more blades means that the machine is dealing with more drag.

Two- and three-blade systems aren't the only options. In sailboat-size turbines, you may find machines with half a dozen or more blades. And from a design standpoint, you can make a case for one blade being the best. This design presents the least drag to the machine, and the least turbulence to following blades. The problem becomes balance — these machines need a counterweight. This design has been produced, even on large-scale machines, but no manufactured machines fit this description today.

I suggest you look at what the market has decided. Most wind generator designs available today are three-bladed because that's what's proven to work best. Think of it this way: Almost all cars today have four wheels — you don't see any to speak of with three, five, or seven. This trend isn't because some genius design has been suppressed by some evil cabal but because four wheels works best. The same principle applies to wind-generator blades — most machines are three-bladed because that works best. Others have two blades or more than three, and making them work well is possible but more challenging or costly.

Upwind versus downwind

The upwind versus downwind distinction basically tells you which side of the tower the wind is coming from (see Figure 13-5):

- ✔ **Upwind:** An upwind turbine has its blades upwind (surprise!) of the tower. That is, if the wind is blowing from the south, the blades of the turbine are on the south side. When the wind shifts to the east, the turbine yaws (turns) around to face east. This configuration is the most common configuration for all scales of turbines, from sailboat- to utility-scale (see "Surveying Wind Generator Sizes" earlier in the chapter for more on turbine scale).

- ✔ **Downwind:** A downwind turbine has its blades on the downwind side of the tower. If the wind is blowing from the south, the blades are on the north side of the tower. And when the wind shifts to the east, the blades move around to the west side of the tower. You often see downwind turbines in home- to small utility-scale turbines.

The distinctions between the two configurations aren't compelling — they aren't factors in my buying decision. No one I know in the small wind industry would say, "You should buy an upwind turbine," or "You should buy a downwind turbine," unless it's an aesthetic preference. Good advisors will say, "You should buy a reliable, productive turbine." Neither of these designs shows a prominent difference in performance or reliability over the other, so when I explain the following advantages and disadvantages, take them as minor points:

✔ Upwind turbines must have a tail or other means to direct the rotor into the wind. This factor does mean extra cost and complexity. Downwind turbines' rotors generally orient themselves properly to the wind without outside assistance. Some downwind designs can occasionally get stuck upwind, which is certainly a drawback but not typically a large one.

✔ *Tower shadow* refers to the effect of the tower on the quality of the wind passing by it. Because of blockage to the wind caused by the tower, a turbine's blades see a different level of wind every time they pass the tower. This effect is more pronounced with downwind turbines because the tower sees the wind before the blades do. But again, this isn't a major problem.

✔ Downwind turbines have a slight advantage in avoiding *tower strikes* — situations where the blades actually hit the turbine, usually resulting in catastrophic failure. Tower strikes rarely happen with downwind turbines because the wind is pushing the blades away from the tower. Upwind turbines just need to be designed with appropriate clearances so that the (sometimes significant) bending of blades doesn't lead to disaster.

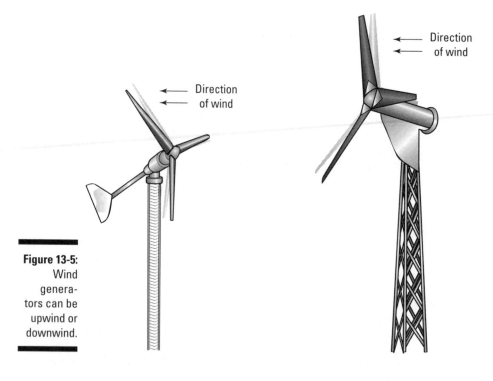

Direction
of wind

Direction
of wind

Figure 13-5:
Wind
genera-
tors can be
upwind or
downwind.

Upwind

Downwind

Understanding blade chatter

Pretend you're a wind turbine facing the wind. Extend your arms straight out sideways from your shoulders, horizontally. Imagine that the wind changes direction, and you have to *yaw* — that is, turn around the tower to face the changing wind direction. If your blades (arms) are extended out with your hands flat to the wind, you encounter resistance to yawing — the flat of the blade is trying to push against the wind, which makes yawing hard. More important but less intuitive is that there's higher inertia (the resistance to a change in motion) in blades when they're extended to the sides and less when they're extended up and down.

Now put your blades (arms) up and down, with one over your head on a slight angle, and the other down by your opposite leg. Now try yawing. You notice that with your palms flat, the edges of your hands easily cut through the wind as you yaw.

Now imagine that you're going from hard-to-yaw to easy-to-yaw several hundred times a minute. Are your shoulders starting to hurt yet? You've just experienced the beginnings of blade chatter.

Only a few current models of downwind turbines are on the North American market, and they can be fine choices for your wind-electric system. I don't recommend seeking out turbines strictly for this feature, but I also don't recommend avoiding them because of it.

Permanent-magnet alternator, wound-field alternator, or induction machine

Here's a quick science lesson for you: Making electricity with rotating equipment amounts to moving magnetism past wire. Copper is a very conductive material, so in a wind-electric system, you're generally moving magnetism past copper wire. The magnetic field induces a voltage in the wire, which moves charges through the circuit, moving energy from the alternator to the electrical *load* (your appliances, lights, and other electricity users).

The primary types of alternators used for wind generators can be defined by how they make the magnetism (see Figure 13-6):

  ✔ **Permanent-magnet:** These alternators use *permanent magnets,* which have a permanent and constant magnetic force. These magnets are either ferrite or neodymium (very strong magnets made with *rare earth* — a mineral compound found primarily in China). If you pass a permanent magnet over a copper wire or coil, you make electricity.

✔ **Wound-field:** These alternators use electromagnets rather than permanent magnets. *Electromagnets* are based on the fact that a conductor that has charges flowing has an electric field around it. If you wind a copper coil and then run electricity through it, you get a magnetic field around the coil (hence the term *wound field*). In this system, you're actually passing copper coils (field coils that are creating magnetism) past other copper coils that generate the end-result electricity of the alternator.

In terms of advantages and disadvantages, these two magnet styles have no major pluses and only minor drawbacks. Wound fields have the advantage of matching the magnetism (and therefore power) with the wind speed. However, they have to use a dab of the source energy to get the magnetic field going. Permanent magnets don't have this small loss, but their magnetism is constant, so it's theoretically perfectly matched only to the wind energy on the blades at one wind speed.

A third class of wind generators is induction machines (refer to Figure 13-6). These use an induction motor/generator that actually syncs to the utility grid without the need for an inverter (the electronic device that converts the DC output of most wind generators to grid-synchronous AC). The magnetic field is initiated either by capacitors (electronic "batteries" of sorts) or by the grid, and the stator delivers a rotating magnetic flux, which induces amperage from the coils in the rotor. These generators are very common in utility applications, utility-scale wind turbines, hydro turbines, and other applications. They're often motored up to speed, and then the wind pushes them a bit faster, reversing the charge flow so they're making energy instead of using it while remaining synced with the grid.

Although this type of machine has been used for many years on various sizes of turbine, it hasn't been so common in the United States of late. The end result is that inspectors and designers are less familiar with this design and may not specify or approve it. In fact, it's as safe as its permanent-magnet cousins and is a very effective way to grid-tie a turbine because it reduces the amount of electronics (voltage clamp and inverter) the system requires. It makes a very simple and cost-effective batteryless grid-tie system.

Because it connects to the grid directly rather than via an inverter approved for grid connection, it's necessary to include certain electrical safety controls that would normally have been built into the inverter.

So which kind of alternator do you want? This usually isn't a buying decision point, which is good because you don't have a lot of choice: No single manufacturer gives you a choice between the three types. A few older machines on the market use wound fields, while most use permanent magnets, and a few of the newer home-scale designs are induction. In the end, your choice of wind generator probably won't rest on the alternator type.

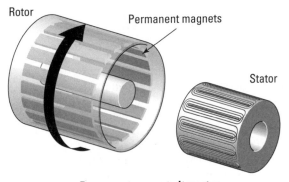

Permanent-magnet alternator

Wound-field generator

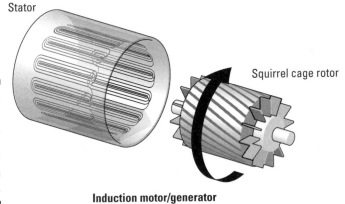

Figure 13-6:
Permanent-
magnet,
wound-
field, and
induction
generators.

Induction motor/generator

Choosing a Wind Generator in the Real World

You can spend years becoming an expert in all the facets of wind generators and wind-electric system design and installation. My bet is that you don't plan to do that, but you still want to own a system. And you're reading this book because you want to make wise choices — I applaud you. Gaining some familiarity with wind generator configuration will help you understand specifications and sort out the important decisions from the less important. But in the end, I find that the technical distinctions I discuss earlier in this chapter are often less important than larger design issues, manufacturer reliability and service orientation, and matching the machine to the end user.

This section provides an overview of the process that you and your partners should go through to make your wind-electric generator choice. (For more detail on the complete design and purchasing process, see Chapter 16.)

After reading this book, you may say that you know enough to be dangerous. Well, you may also know enough to choose wise partners for your project. Find people you can trust and quiz them carefully. They can't make you an expert, but you can tell whether they're on your side and acting as your advocate. Chapter 12 has pointers on gathering a team of experts to help you.

Site and wind resource evaluation

First and foremost, evaluate your site before choosing a wind generator. Your wind resource is the most important factor in the success of your wind-electric system. Don't ignore this step. Everything depends on it. If you don't know what your wind resource is, you don't really know much of anything about how effective a wind-electric system may be for you. See Chapter 8 for more information on what to look for yourself or ask for from a professional.

With the following three pieces of information in hand, you're well on your way to deciding what you need and therefore what you should buy:

 ✔ **Proposed tower site:** This location is often on the highest point on your property, or at least a high point of some kind. It depends on distance to your home and distance to the utility grid — if the highest point is right next to the house, you'll minimize the cost of transmission. You want the tower to be away from canyon walls or rising hills and a reasonable distance from the existing electrical infrastructure (though "reasonable" may be more than 1,000 feet depending on the scale and budget of the project). In choosing your tower site, you're also moving toward the goal of determining your actual wind resource. (Check out Chapter 8 for tips on choosing a site and estimating your wind resource.)

✔ **Tower height:** Tower height is a huge question with wind-electric systems, and choosing the wrong answer (too short of a tower) has negative impacts for the life of the system. Site your tower at least 30 feet above anything within 500 feet, and plan for tree growth. Ignore this rule of thumb at your kilowatt-hour meter's peril! (Flip to Chapter 14 for general information on towers.) Determining your tower height is the next step toward determining your resource — you have to have an average wind speed to work from before you can determine how much of a wind generator you need.

✔ **Average wind speed at the tower height:** This figure is crucial to making reasonable estimations of the energy production, which is crucial to choosing a machine that can satisfy your goals. Apply your wind speed numbers to the manufacturer's predictions, to online calculators, and to formulas (see Chapter 16) to estimate energy production for various turbines.

Your average and peak wind speeds also determine whether you need a very durable turbine or whether a medium- or light-duty machine will last on your site. Overall, I recommend getting the toughest machine you can afford, but you certainly want to avoid putting a light- or medium-duty wind generator on a heavy-duty site. If your average wind speed is above 10 mph and your site sees winds at 60 mph or more on a regular basis, you don't want to buy anything but the toughest machine.

Your average wind speed also helps you define the size of the wind generator you'll buy. Look at energy predictions for various machines at your projected average wind speed at hub height. You should make estimations of wind speed at a variety of hub heights above the minimum to exceed the 30/500 rule (see Chapter 14). Chapter 16 helps you walk through this process.

Owner evaluation

After you evaluate your site, the next step is to evaluate yourself. What type of person are you? Do you want to do everything for yourself, learning about each step and gaining the skills to accomplish them? Or do you want to hire everything out and not touch a nut or bolt?

Knowing your interests, abilities, and limits helps you decide which approach to take when you specify your system components and your installation, operation, and maintenance plan. It also affects what type of wind generator you buy and which specific machine you choose:

✔ If you're the hands-on type and want to do all the work yourself, your options are open. You may fruitfully choose any system type, because you're apt to maintain batteries and the wind generator well and tackle any problems that surface.

✔ If you're the hands-off type, I suggest a batteryless system and simple, reliable equipment that requires only easy maintenance. You need to make a financial commitment to regular maintenance and necessary repairs, or your investment in wind energy will become a loss, not a gain. Taking on a battery-based system or a less-robust wind generator requires more attention, which in your case means more money for outside maintenance on a regular basis. You can decide to go this way, but soon you'll lose any money you save upfront on a cheaper system.

My advice is to buy the simplest, most robust machine you can buy, unless you're really looking to get out of the house and spend more time on your tower. No one regrets buying a machine that gives them less trouble. Many regret buying machines that give them more trouble than they hoped for.

Budget evaluation

What's your budget for your wind-electric system over its lifetime? This question is crucial to answer before buying equipment. In general, wind energy is a very bad place to buy cheap. The sweetness of a low cost upfront soon turns to sourness over the long-term cost and trouble. Follow the wisdom of second-time wind generator buyers, who tend to look at the more-expensive machines because they've discovered that cheaper isn't better or even cheaper over time.

I cringe when I hear people say things like, "That wind generator is too expensive." They're likely looking only at the upfront cost, which is just one part of the picture. It's hard to put a value on a reliable turbine that keeps producing year after year and far easier to find people who can catalog the costs of repeatedly repairing machines that look less expensive upfront. (For tips on calculating long-term costs, head to Chapter 10.)

When looking at the costs, be clear about the whole *system* cost. A wind generator is useless without a tower, wiring, controls, and all the other system components. And these extras all add up in cost. If you don't have the dough to buy well and buy complete, maybe wind energy isn't the best choice for you. Although wind turbines in the home-scale class may range in cost from $2,000 to $40,000, the *system* cost ranges from $20,000 to $200,000 or more. Know what you're getting into upfront instead of buying a wind generator on impulse and finding that you can't afford the components that make it work.

Your wind generator may be anywhere from 10 to 30 percent of the whole system cost. But you're buying a *system* and should look at whole system cost. Don't get fixated on the wind generator and ignore less glamorous but equally important supporting components.

Putting it all together

Choosing the right wind generator can be an involved process. Looking at it as a series of choices may simplify your decision:

- ✔ Do you need a small or large amount of energy?
- ✔ Is your average wind speed high or low?
- ✔ Are your peak wind speeds intense or modest?
- ✔ Are you off-grid or on-grid?
- ✔ If you're on-grid, do you want outage protection (battery backup)?
- ✔ Are you hands-on or hands-off?
- ✔ Is your budget limited?

If you ask yourself these questions and seek honest answers, your wind generator buying decision isn't hard. With the answers in hand, you probably face a choice between two to four machines. At that point, leaning on your partners and getting feedback from actual installers and users of all the machines you're considering should narrow the choice down to your machine.

Chapter 14

Talking about Towers

A tower is a crucial part of any wind-electric system. Without a tower, your wind generator is like a shovel without a handle — perhaps strong and attractive but not functional.

Important factors come into play when you're choosing a tower for your wind system. For instance, tower height is vital to good performance of a wind turbine. The most common mistake for first-time wind system buyers is buying and installing too short of a tower. After you decide on tower height, you need to decide which tower type is best for your situation. Your choices are tilt-up, fixed guyed, and freestanding.

This chapter helps you make crucial decisions about your system's tower, saving you time, money, and frustration down the road.

The Air up There: Tower Height

When you're designing a wind-electric system, you must make two critical decisions that together have an enormous effect on the performance of your system. The first is the size of the rotor, which I cover in Chapters 5 and 13. The second is tower height.

Choosing a tower height is a somewhat irrevocable decision. You can't just add more sections unless the whole installation — from the foundation up — was designed with that strategy in mind. So make a wise decision upfront and buy tall. In this section, I explain the importance of tower height and give you pointers on selecting the best tower height for your system.

If you want to cut to the chase and avoid the tech talk, just buy a taller tower than you ever imagined needing. Seasoned dealers use the *30/500 rule* (in which the lowest blade tip is at least 30 feet above anything within 500 feet) as a minimum starting point, but taller is better. You won't regret it. In all my years in the small wind industry, I have never heard anyone say, "I wish I'd bought a shorter tower." But I repeatedly see situations where people wish they'd gone taller.

Understanding how height influences power output

Your goal in installing a wind-electric system is to generate electricity — measured in kilowatt-hours (kWh). And the fuel is wind. The more fuel you have, the more electricity you get.

You find steadier, stronger winds the higher you go in the sky, so putting your wind generator up in the best fuel possible is important. You wouldn't buy a solar-electric module and put it in the shade, and putting a wind turbine on a short tower, where the wind is weak and very turbulent, is just as pointless. In the following subsections, I explain why the wind is stronger and less turbulent (and therefore much more effective for your wind system) when you're higher off the ground. I do so with the help of two important concepts: wind shear and wind speed cubed.

Wind shear

Wind speed increases as you move away from the ground, which has a rough surface of buildings, trees, hills, and gullies. The rate of wind speed's increase is called *wind shear,* and it's described by a mathematical coefficient. I'll spare you the math, but the bottom line is that if you're in rough terrain, the wind speed increases very quickly as you get higher; if you're over smooth terrain, it increases more slowly as you get higher. Various sources divide terrains into different classes, or "roughnesses," and they generally look something like this, from smooth (low shear) to rough (high shear):

- ✔ Perfectly smooth water
- ✔ Flat grassland or low shrubs
- ✔ Trees or hills, with buildings in the area
- ✔ Close to trees or buildings
- ✔ Surrounded by tall trees or buildings

Although tall towers are crucial everywhere, going higher is even more beneficial if you live around lots of trees, buildings, hills, and other obstructions. Most people live near aboveground buildings, and humans generally like to have trees around them. These tendencies mean that most people live on sites that have high wind shear. However, if you're in the flat plains of Oklahoma in an underground house and you keep your lawn mowed short, the wind shear is lower, and a really tall tower isn't as vital (although I still recommend going higher than you think you need).

In Chapter 16, I show you how to use this wind shear information in calculating your average wind speed at hub height. For now, it's enough to understand that tall towers are important wherever you are, but they're even more important if you live with trees, houses, and other obstructions around.

Wind speed cubed

The power available in the wind is based on the wind speed cubed (wind speed $\times$ wind speed $\times$ wind speed, or V^3). This isn't like the linear relationships that you're probably used to. If you give your car's engine twice as much throttle, you expect to get about twice as much speed. However, the relationship between the wind speed coming in and the energy coming out is cubic rather than linear. That means that if the wind speed doubles, you stand to get eight times as much energy output.

Because wind speeds increase the farther you get from the ground, you can significantly increase your wind power just by putting your wind generator on a taller tower. Small increases in wind speed mean large increases in electricity output. For example, going from 10 miles per hour to 12 miles per hour doesn't result in the 20 percent increase in wind energy you may expect — it's a 75 percent increase. You may see that sort of increase by adding as little as 40 feet of tower height, depending on your site. Flip to Chapter 5 for more details on wind speed cubed as it relates to the power of the wind.

Selecting the right height for your tower

Choosing your tower height is not a simple, linear process. If your goal is to generate enough energy to power your home, you need to know your average wind speed (see Chapter 5). But tower height affects average wind speed, so you need to make a preliminary determination of tower height to get the average wind speed. Then you can home in on the other factor in how much energy you'll get: swept area (see Chapter 13). And after you've made a preliminary determination of which wind generator you'll use (based on its swept area and/or the manufacturer's predictions), you may decide to increase your tower height to increase your energy output.

When choosing your tower height, make sure you look at possibilities taller than your initial plan. Suppose you decide that an 80-foot tower is all you can afford. Well, price out 100-foot and 140-foot towers as well, and make an estimate of the wind energy available and the energy production predicted at those heights (see Chapters 5 and 8 for the principles and Chapter 16 for practical steps). In many cases, you find that you can't really afford *not* to go higher if your goal is to produce wind electricity economically.

For example, if an 80-foot tower costs $6,000, a 100-foot tower costs $8,000, and a 140-foot tower costs $12,000, you need more information to make an intelligent buying decision. The primary question is how much more energy you get at a taller height (say, 140 feet) versus a shorter height (such as 80 or 100 feet). Looking at the wind shear on your site, a wind energy expert can use your estimated or measured wind speed to come up with an estimated average wind speed for a different height.

What you find is that unless you live on a site with very few obstructions, going higher is worth your while in most cases. Rarely does going higher not result in increased energy production. Limitation on tower height may come from your overall budget and perhaps from legal restrictions (see Chapter 2), but the economics generally push you to taller towers, within the range of choices most people have.

My minimum tower height for an underground home in the prairie is 60 feet. Typical rural system towers in open country may range from 80 to 150 feet, depending on the terrain and obstructions. **Remember:** The rule of thumb is to have the lowest blade of the wind generator be at least 30 feet above anything within 500 feet.

Depending on the style of tower you use (I describe different styles later in this chapter), you may or may not have many choices when it comes to height:

- ✔ Tilt-up towers come in specific heights from specific manufacturers. For example, Abundant Renewable Energy in Oregon offers tilt-ups that are 43 feet, 64 feet, 85 feet, 106 feet, and 127-feet tall. Other manufacturers have other increments.

- ✔ Guyed lattice towers (which are the most common type of fixed guyed tower) and freestanding towers generally come in 10- and 20-foot sections, giving you great latitude in tower height choice. Each different height requires specific engineering and specifications for foundation and guys.

Examining Types of Towers

Three different tower styles are on your menu when you're specifying system components: tilt-up, fixed guyed, and freestanding. Your system, site, budget, and personal preferences determine the choice you make. Here's a quick breakdown of all three towers (Figure 14-1 compares their *footprints,* how much cleared space they take up on the ground):

- ✔ *Tilt-ups* lend themselves to open sites and convenient, on-the-ground maintenance.

✔ *Fixed guyed towers* are the most common, least expensive, and most adaptable to different sites.

✔ *Freestanding* towers are the safest, most expensive, and perhaps most aesthetically pleasing.

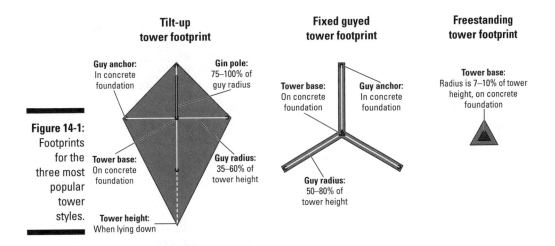

Figure 14-1: Footprints for the three most popular tower styles.

Read the following subsections for details on these towers (and a few alternatives you should avoid). Then make your choice based on your site, preferences, and situation, as I show you how to do later in this chapter.

All tower costs I mention in this section are rough current estimates in U.S. dollars. Tower prices regularly change depending on the market, the cost of steel and transportation, and so on. Generalizing and making comparisons is difficult because there's a lot of variety across tower types, sizes, and manufacturers. After you've determined what wind generator you plan to buy, get quotes for a number of towers in the types you're considering.

Towers are described with some specific terms that may not be familiar to you, so here's a mini glossary (see Appendix A for even more terminology):

✔ **Tubular:** Tower sections that are made of pipe or tube

✔ **Lattice:** Tower sections that use steel tubes or rods for verticals with horizontal and diagonal bracing

✔ **Guy wire or guy:** Steel cable that supports the tower

✔ **Guy anchor:** Concrete or other material holding the guy wires into the ground

✔ **Tower base:** Concrete or steel foundation directly underneath the tower

Tilt-up towers

For safety, convenience, and serviceability, tilt-ups are strong contenders. If you have the right site for a tilt-up tower and a tilt-up tower is available for the wind generator you have, it's likely your best choice.

Tilt-ups can be tubular or lattice and are engineered to be raised and lowered with a winch, grip-hoist, or vehicle and a system of cables and pulleys (see Figure 14-2). These towers have four sets of guy wires with four anchors, a tower base, and a *gin pole,* which is a lever that allows you to tilt the tower using cables and pulleys. The tilting allows you to perform turbine work on the ground. Though the first few lifts may be a little challenging (and I recommend expert support for those), you get into a routine after you've done it a few times, and you can do the lowering and raising very safely and quickly.

In this section, I describe different kinds of tilt-up towers, discuss the pros and cons of tilt-ups, and explain the costs involved.

Kinds of tilt-ups

Most tilt-up towers on the market are of the tubular variety. They consist of sections of tubing or pipe connected with couplers and supported by guy wires (the tower in Figure 14-2 is tubular). Sometimes the tower kits are sold without the pipe or tube, so you can buy locally and save on shipping costs.

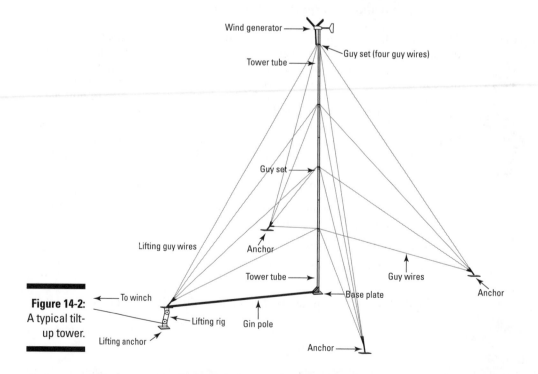

Wind generator

Guy set (four guy wires)

Tower tube

Guy set

Lifting guy wires

Anchor

Tower tube

Guy wires

Anchor

Figure 14-2:
A typical tilt-up tower.

To winch

Base plate

Lifting rig

Gin pole

Lifting anchor

Anchor

Occasionally, you find tilt-up towers made with lattice sections similar or identical to guyed lattice tower sections. These models have the advantage of being climbable after you tilt them, but they're more expensive.

The advantages

My colleague Mick Sagrillo is fond of saying that some people think that if God had wanted them to be 100 feet in the air, he would've given them 95-foot legs. If that describes you, a tilt-up tower may be just the ticket. Here are a few of the benefits of tilt-ups:

- ✔ The biggest advantage of these towers is that you can do all the work on terra firma. No climbing gear is necessary; no climbing fear needs to be overcome. Anyone who has done a lot of tower work understands from experience that everything takes longer up there — maybe three to five times longer. When you have to be concerned about a dropped bolt or a dropped wrench, you have to slow down. With tilt-ups, folks who are less physically able can still be do-it-yourself wind system operators.

- ✔ With tilt-ups, you don't need to be as concerned about the weather. Whereas you'd think at least twice about climbing a tower in very cold or wet weather, you can pretty easily (if not pleasantly) work on a turbine in these conditions with a tilt-up — though you still don't want to raise or lower these towers in heavy winds.

- ✔ In as little as 30 minutes, you can have the tower rigged and lowered. So if you live in hurricane country, you can tilt your investment down before the storms. After it's lowered, realizing that you've forgotten a tool or bolt doesn't mean a trip down the tower and back but rather a short walk to your truck or shop.

The disadvantages

Despite their many advantages, tilt-up towers do have a few negatives:

- ✔ The biggest disadvantage of tilt-up towers is their footprint — they take a lot of cleared space on the ground (refer to Figure 14-1). The space required for a tilt-up is a large diamond shape, with the distance from the center to one point equal to the tower height and the other three distances equal to the guy wire radius. With taller towers, this space can be a large area: For instance, a typical 120-foot tower requires a diamond shape of about 180 feet by 120 feet. Within most of this diamond, you can't have anything taller than high grass because all the guy wires need to be free to lie down.

- ✔ Tilt-ups need reasonably level ground. I've seen tilt-ups installed on very uneven sites, but ideally you want to have level ground across the two side guy anchors and base.

If you're lifting and lowering on unlevel ground, watch out for variable tension in the guy wires. Loose guy wires can mean a sloppy, noodly tower, and overly tight guy wires can actually collapse a tubular tilt-up.

✔ Tilt-ups are typically more expensive than comparable non-tilting fixed guyed towers (which I discuss later in this chapter). They involve more hardware, with four sets of guy anchors and wires rather than three. This extra equipment also means that assembly and inspection take longer.

✔ Though not having to climb is an advantage, not being able to climb can be a disadvantage. Sometimes a minor repair or check is in order, and a climbable tower would mean a quick trip up to inspect or repair. With a tilt-up, no matter how minor the problem, you must tilt the tower down.

✔ Compared to freestanding towers, tilt-ups have the disadvantage of having guy wires. These are an aesthetic cost and a maintenance expense. Guyed towers are also more subject to failure than freestanding towers.

The costs

Tilt-up tower costs vary depending on the rotor diameter of your generator; specific machine and tower style; height; *frost depth* (how deep your foundation needs to be to get below frost heaving), which equates to how much concrete you'll need to buy; and other factors.

A 100-foot tilt-up tower kit for one 8-foot rotor machine on the market today costs about $2,700. For one 12-foot diameter machine, a 106-foot tilt-up tower costs $5,500. These costs are for the tower kit, which typically includes tube or lattice, guy hardware, anchors, and all associated hardware. They don't include excavation for the base and anchors, concrete, or labor, which add several thousand dollars.

When analyzing the cost of a tilt-up tower, compare apples to apples and account for all costs. Although tilt-ups may be somewhat more expensive than fixed guyed towers (which I describe later in this chapter), if your alternative is to hire a crew to do maintenance on a climbable tower, the tilt-up comes out ahead.

Fixed guyed towers

Fixed guyed towers are very versatile because you can use them on a variety of site types and adapt them for a wide range of heights. Guy wires are attached at intervals to keep the usually lattice (but occasionally tubular) tower upright — without them, the tower would fall (see Figure 14-3). These towers don't tilt up and down. Because they're widely used, they're easily available, as is expertise for their design and installation.

In this section, I describe different kinds of fixed guyed towers, discuss their pros and cons, and explain the costs you may encounter.

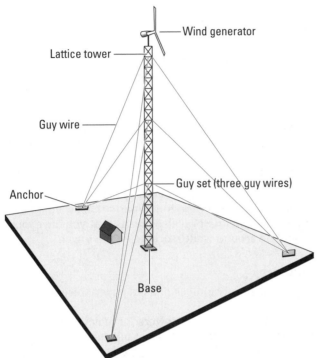

Types of fixed guyed towers

You can find two basic types of fixed guyed towers in the field today — lattice and tubular:

- **Lattice:** Lattice towers are far and away the most common. These are the towers you see at your local radio station or in your local amateur radio enthusiast's backyard. They come in 10- or 20-foot sections, and they have either tubular or solid steel legs. They're guyed on three sides (at 120-degree intervals around the tower) and have guy wires somewhere between every 30 and 80 feet up the tower, depending on the design. (Figure 14-3 depicts a lattice tower.)

  In telecommunications applications, guyed lattice towers are used to heights beyond 2,000 feet, which dwarfs the heights of towers used for wind generators.

- **Tubular:** Fixed guyed tubular towers are unusual but a reasonable option if you can find one or want a build-it-yourself tower. These setups use sections of pipe welded to each other to form a continuous tower tube, sometimes starting larger and tapering smaller as it goes up. But instead of being freestanding, these towers are kept upright with guy wires attached at intervals and descending to the ground in three directions.

The advantages

Fixed guyed towers have several advantages:

- ✔ These towers are frequently the least expensive option for a given wind turbine. Concrete requirements for the base are modest compared to those of a freestanding tower (which I describe later in this chapter), and fixed guyed towers have one fewer set of guy wires than a tilt-up.

- ✔ You can install fixed guyed towers on a wide variety of site types. Uneven ground isn't a liability as long as you can excavate for the tower base and anchor concrete. You can even site these towers in a forest as long as you clear lanes for the guy wires and remove hazard trees.

- ✔ Fixed guyed towers are easy to climb because the tower section construction includes regular horizontal parts that serve well for steps and hand grips. This construction also means that you have lots of things to clip into and tie around while working on the tower.

The disadvantages

Beware the following disadvantages of fixed guyed towers:

- ✔ You have to climb fixed guyed towers for maintenance — you don't have the option to tilt them down. This fact means you need a climbing safety system, harness and other climbing gear (see Chapter 17), and the tools prepared so you can work on a tower. You also need the willingness and ability to climb or to hire someone to do the maintenance for you.

- ✔ These towers have guy wires, which, compared to a freestanding tower, is an aesthetic and practical disadvantage. Fixed guyed towers need three clear lanes for the guy wires to run in, and they're vulnerable to guy wire damage from falling trees.

The costs

Cost for a 120-foot fixed guyed lattice tower kit for a 21-foot diameter machine is about $15,000. This sort of kit typically includes lattice sections, guy hardware, anchors, and all associated hardware. It doesn't include excavation for the base and anchors, concrete, or labor.

Installation cost is generally higher with these towers than with tilt-ups. Both of the two common methods of installation increase the cost. The easiest way is to use a crane, which is an expense to hire. The labor-intensive way, using a temporary *gin pole* (a vertical pole clamped to tower leg and extending above tower) and raising a section at a time, involves a lot of labor. I've erected fixed guyed towers with a crane in one day, but three or four people would need several days to do it without a crane.

Freestanding towers

Many people consider a freestanding tower (see Figure 14-4) to be the crème de la crème of towers. If you want or need a climbable tower with a small footprint, it's likely your best choice. Its primary drawback is cost. Holding a big stick topped with a large pinwheel up in the air without guy wires requires lots of steel and concrete!

In this section, I describe different kinds of freestanding towers, go over their pros and cons, and note the costs you face.

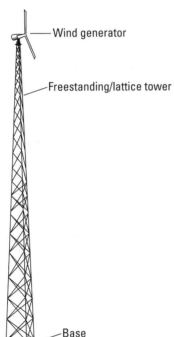

— Wind generator

— Freestanding/lattice tower

— Base

Figure 14-4:
A
freestanding
tower.

Different types of freestanding towers

Two standard types of freestanding towers are on the market today — lattice and tubular. Availability depends on the height you desire and the rotor diameter of the turbine.

　　✔ **Lattice:** Lattice towers look somewhat like the Eiffel Tower, typically having three pipe legs joined by angle-iron bracing. (The tower in Figure 14-4 is a freestanding lattice tower.) Each section tapers, starting 10 to 20 feet apart at the base and ending up perhaps 1 or 2 feet apart at the top, depending on the height and size of turbine. You climb lattice towers by scaling climbing rungs on one of the legs while wearing a harness attached to a safety cable and fall-arrest device.

✔ **Tubular:** Freestanding tubular towers (often called *monopoles*) usually appear on community and utility-scale turbines and are becoming more popular for home-scale turbines (head to Chapter 13 for more on turbine size). These towers are steel, though I've heard rumors of the possibility of fiberglass as well. Many people prefer the aesthetics of these towers — you can paint the smooth columns however you want. They start with diameters in the 2 to 4 foot range, depending on the height and machine carried, and they have climbing pegs built into one side. Tubular freestanding towers are generally more expensive than lattice towers.

The advantages

Freestanding towers appeal to people who want a minimum of visual intrusion. They work well for tight sites, and are the safest tower type to install. Here are the advantages of freestanding towers:

✔ Chief among the advantages of this tower style is its very small footprint (refer to Figure 14-1). You can install even a 160-foot tower with an excavation area of only 20 by 20 feet and a final tower footprint that covers a triangle with 16-foot sides. This small space requirement means that you can install this tower in very close quarters — even in a backyard in the suburbs, if you can get permission.

✔ You can install freestanding towers on uneven sites that can be challenging for fixed guyed towers and nearly impossible for tilt-ups. Freestanding tower sites need access for excavation equipment and cranes, but they can be installed in very tight situations. One wind-electric installation in my neighborhood has a 400-foot cliff dropping off about 80 feet from the tower and a hill rising on the opposite side. With a very small clearing and a driveway entering near the tower, a guyed tower would be extremely difficult to install on this site.

✔ Freestanding towers are perhaps the safest of all towers to install because they're installed with cranes. If you have space to lay the full tower out on the ground, you can completely assemble it, install the wind generator, run the conduit and wire, and then lift the whole tower as one piece. The only work necessary on the tower is to climb up and remove the crane strap after you secure the tower to the foundation. These towers also are generally the strongest and safest in severe weather because they don't rely on vulnerable guy wires to stay upright.

✔ Freestanding towers have no guy wires, which gives them aesthetic and practical advantages. My wife once quipped that we have so many guy wires around our house that we may as well be on the grid. Guy wires are visible at close range (though they disappear when you get a few hundred feet away). In addition, guy wires require guy anchors and cleared areas for the guy wires to run in, and these considerations can disrupt building and landscaping plans. Freestanding towers avoid these concerns.

The disadvantages

Take a look at your pocketbook, your environmental values, and your physical abilities as you analyze the disadvantages of this type of tower. Most people who can afford freestanding towers find that the advantages outweigh the disadvantages.

✔ The number one disadvantage of freestanding towers is cost. They can easily end up being twice the cost of a fixed guyed tower of the same height and rotor diameter. This discrepancy is due to the amount of concrete and steel required to hold a lever arm of that height up in storm winds. A common 100-foot tower I've installed requires 40 to 50 yards of concrete in the ground and lots of steel. Most of the steel is above the ground — tubular legs and angle iron braces. But significant steel reinforcement and anchor bolts also are in the concrete. These bolts can be upwards of $1^{1}/_{2}$ inches in diameter and 6 feet long to tie the tower steel to the concrete base.

In addition, you need a crane for installation, making these towers more expensive than tilt-ups (though the crane time can be very short if preparation is done well).

✔ A related disadvantage is the environmental cost. Concrete is very energy intensive, as is steel. They're both long-lasting materials, which tempers this problem somewhat. But ask yourself how your environmental goals are served if you have to use tons of concrete and steel to make renewable energy.

✔ You have to climb most freestanding towers for maintenance (though a few manufacturers of very short freestanding towers have designed tilting ability in). To the aerially inclined, this fact is an advantage, not a disadvantage. But it's a requirement nonetheless. As with fixed guyed towers, you need the appropriate safety equipment and tools for work aloft (see Chapter 17). And for those of you who are nervous up there, freestanding towers shake more than guyed towers — fair warning.

The costs

A 120-foot freestanding lattice tower for a common 21-foot diameter turbine costs $18,400 for the steel tower kit. The same height in a tubular tower for the same machine costs $41,800. In both cases, these numbers don't include tower shipping, excavation for the base, concrete, or labor.

Steering clear of alternative towers

You may be tempted to do something "creative" to get your wind generator up into the wind. I advise caution. Though dedicated homebrewers (including me) have been successful with some creative ventures some of the time, a majority of experimental towers and mounting systems either do not work well or do not survive for the long term.

Roof mounting

Chapter 8 discusses turbine siting at length, and the beginning of this chapter focuses on the critical nature of tall towers for successful wind-electric systems. So I hope you're not surprised that I strongly recommend that you do not attach a turbine to a building in any way, shape, or form.

This classic hare-brained wind idea is one that you should run, not walk, away from. Why? Wind generators need wind. Where is the wind? High above ground obstructions — the standard guideline is that the lower tip of a wind generator should be at least 30 feet above anything within 500 feet. This is quite difficult to attain when you start on a roof. But even if you can figure out how to do it, don't! Here are just a couple of reasons why:

- ✔ The closer you are to buildings — vertically and horizontally — the less wind and more turbulence you have. So even if you obey the 30/500 rule, the best site on your property is still probably not right next to your house. You want to examine the prevailing wind patterns and put your turbine in the smoothest and strongest flow you can find.

- ✔ Wind turbines spin, and anything that spins has vibration. Vibration means noise, and buildings can act as *soundboards,* magnifying the sound just like the top of a guitar lets you hear the strings.

  In addition, vibration has a way of moving things. One Colorado woman was the victim of a fly-by-night installer who stuck a 10-foot diameter turbine just above the peak of her garage roof, attached to the ridge. In her open prairie area, it did run some — loudly and roughly. A few years later when a pro installer came to her rescue, he found the nails in the garage's rafters had started to vibrate out. If left in service, the turbine would've deconstructed her building!

Sensible design rules out turbines attached to buildings, because sensible design includes a resource assessment. A quality *resource assessment* looks at the available wind in various locations on the property and at various sites. Even rudimentary research techniques show that the wind around buildings is weak and turbulent compared to the wind well above buildings.

So why do people pursue this dead end? I think it's largely an attempt to cut costs. Cutting upfront costs isn't a good way to approach wind-electric systems; it's actually a recipe for poor performance. One study, the Warwick Wind Trials in Great Britain, found that the urban resource is very poor indeed. Based on those figures, I roughly calculate that the cost of energy (COE) for the small roof-mounted turbines in that study was about $1.50 per kilowatt-hour if you assume (generously) a 20-year lifespan for these turbines, with no maintenance costs (both unrealistic, but I'm bending over backward to be fair). This number compares with 3 to 25 cents per kilowatt-hour for electricity from the utility grid.

In short, roof mounting is a bad idea that wastes your time and money, regardless of what the many do-it-yourselfers and scam wind turbine companies pushing this idea try to tell you. Roof mounting simply defies the laws of physics and economics. Put up a proper tower clear of all nearby obstructions, and your wind turbine will perform well for you; cut corners, and performance will suffer.

Homebuilt towers

Wind people get crazy ideas in their heads. They want to build their own turbines, often with strange and wonderful designs. They also sometimes want to build their own towers. Home-building a tower is risky business that you should approach with *extreme* caution.

What can happen if you try to build your own tower? When you become the engineer, you take responsibility for the structural integrity of the tower design. Although I'm all for independent experimentation, it's worth noting that engineers study for years to understand the physical dynamics of towers and to be able to calculate the forces. Even they don't always get it right — I could tell you more than one story of engineered tower failures. And having experienced one homebrew tower failure myself, I advise extreme caution in this realm.

I remember climbing one of my homebrew towers with a friend and colleague who's a tower engineer. I took the opportunity to ask him what he thought — whether it would hold up to the storms over time. He raised his eyebrows, gave me a look, and said, "You want me to do some engineering for you?" His point was well taken. I and only I have the responsibility if I build my own tower. Without engineering calculations, I'm making an educated guess, and no engineer — even a friend — is going to want to put his stamp of approval on it without doing the work to verify the quality of the design.

My first advice to you on home-building towers: Don't do it. If you insist on building your own tower anyway and you have extensive experience with metalworking, please heed my second piece of advice: Carefully examine manufactured towers similar to the type you want to build and then overbuild a steel tower. This advice isn't foolproof, so my final piece of advice is not to hold me responsible for the results.

Choosing the Right Tower for You

Choosing a tower's type, height, and location as part of an overall site evaluation are perhaps the three most important decisions in a wind-electric system design. All three have potential for catastrophe, poor performance, aesthetic offense, and disappointment if you don't consider them carefully.

You need to evaluate not only the site but also the owner (you) and the budget if you want to make a good decision. All of these factors interact, and no one should trump the others.

Site evaluation

Site evaluation affects tower type, height, and specific location on your property, as you find out in this section. See Chapter 8 for more detail about the process of doing a site survey.

Relating your site to tower type

First things first: Just think about the topography and vegetation on your property, and consider each type of tower. (Check out the preceding sections for details on the various tower types.) Here are the site requirements for each type of tower:

- **Tilt-up:** A tilt-up tower requires a large open space with almost no obstructions — moving sheep and low bushes are compatible, but high bushes and trees aren't. So if your site is largely wooded with no open fields, you can rule out a tilt-up tower.

- **Fixed guyed:** A fixed guyed tower requires guy-wire access. You can install one of these towers in the middle of a forest, but you have to clear a spot for the tower base, with three lanes radiating out at 120 degrees to each other. If your site doesn't have the space and configuration to work this sort of tower around the buildings, roads, and property lines, you can cut your options down to one — a freestanding tower.

- **Freestanding:** Freestanding towers have a very small footprint, but you do need to have a place where you can excavate an extensive hole without hitting a septic field, plumbing and electrical lines, or impenetrable rock. And you need access and space for a crane to lift the tower.

Determining a workable tower height

Your site also drives your tower height, regardless of which tower type you choose. On most sites, the trees are the tallest things on the landscape, so your first task for coming up with a tower height is to determine the *mature height* (how tall the trees will be in their lifetime, or at least yours) of the tallest species of tree on your property. At the very least, educate yourself to determine how tall the tallest trees will be in 20 or 30 years — a reasonable time frame for a wind-electric installation. If you're a youngster who takes the long view, stick with mature tree height.

After you have that number, add at least 50 feet to it, and you have a pretty good number for your tower height. Earlier in this chapter, I do say to go *30* feet above anything within 500 feet, but 30 feet is a minimum, and it assumes that you're measuring to the bottom of the blade and that you made good estimates of mature tree height. Always round up! You'll never regret having too tall a tower; you'll always regret having too short a tower.

Tower type is another factor in determining tower height. Freestanding towers are not an issue in this regard. Fixed guyed towers require more and more guy-wire radius space as they get taller, and you need to make sure they'll fit on your property and avoid obstructions. As I mention before, tilt-up towers require a large footprint, and the taller you go, the larger the footprint will be. Tilt-up tubular towers are also limited in overall height to about 130 feet, as far as what products are available.

Chapter 16 has full details on sizing your tower height to fit properly within your entire wind system.

Owner evaluation

One of the key decisions that face you when choosing a tower is how you reasonably expect to maintain it. Do you plan to do the work on your turbine yourself? If you are, are you willing and/or able to climb a tower (and if you aren't, is your contractor)? If the answer to any of those questions is no, a tilt-up tower is in your future because it's the only tower that doesn't require climbing. If climbing for maintenance isn't a problem, your options are open. (Flip to Chapter 12 for help with finding a wind-energy professional.)

Your next decision deals with your aesthetic preferences, and I encourage you to include your significant other, kids, other relatives, and especially neighbors in this discussion. Aesthetics are personal. Some people love the look of monopoles and hate guy wires, so be sure to consider the opinions of everyone affected by your potentially monstrous construction. Ideally, you'll visit working wind sites and see different tower styles in the landscape before making your choice. This choice isn't easy to reverse later, so make it wisely.

Budget evaluation

After you've decided on your preferred tower style and the appropriate tower height (or higher), get a few quotes on tower purchase and installation for your preferred style and other possible styles. Then compare those figures to the funds you have available.

Bear in mind that the tower cost is often the largest single cost in the system. Don't try to save money here by going shorter, cheaper, or lower quality or by cutting corners in any way. The tower is the foundation upon which your wind energy collector sits. Investing poorly here will affect everything else in your system, especially its overall performance. Check out Chapter 10 for more guidance on calculating costs.

Putting it all together

Making your tower decision means asking and answering a number of cascading questions. Figure 14-5 features a decision tree to help you.

Whatever you decide, make these decisions as though they're the most important decisions you make in your design process — because they are!

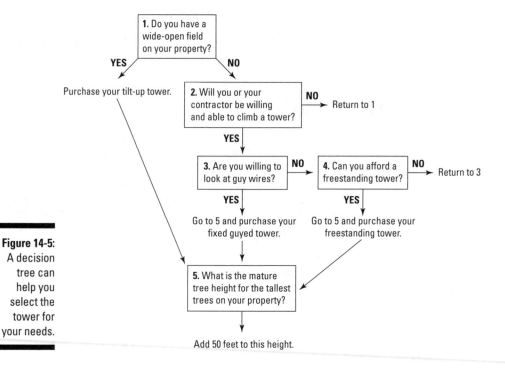

Figure 14-5:
A decision
tree can
help you
select the
tower for
your needs.

Chapter 15

Checking Out the Rest of the System Components

• •

• •

A wind generator and a tower by themselves don't make electricity. All the other components of a complete system are necessary to do that. Understanding the functions of other basic components (often called *balance of systems* [BOS] components) can help you make smart decisions when you're designing and buying a wind-electric system.

In this chapter, I first discuss inverters, which modify the electricity from your wind generator so it has the right type of current and voltage to run your lights or feed the utility grid. If you want a system with battery storage, you can then read about batteries and charge controllers, which prevent overcharging your battery bank (see Chapter 9 for details on systems that use batteries). This chapter wraps up with a discussion of wiring and some key safety components — disconnects, overcurrent protection, and grounding — that help keep you and your equipment from being zapped.

Investigating Inverters: Getting the Right Current and Voltage

Inverters convert the direct current (DC) output of the wind generator to usable alternating current (AC) electricity. Almost all wind-electric systems have an inverter or inverters. The only exceptions are

✔ Direct water-pumping or heating systems

✔ Very small DC-only systems

So unless you fall into those two categories, you have to purchase and use an inverter. The inverter is a key player with a strong part to play. It's not a place to skimp or cut corners — it's a primary component of a wind-electric system. In this section, I explain what an inverter does and describe different types of inverters. I also discuss system voltage as it relates to inverters.

Defining an inverter's main functions

At its root, an *inverter* is a device that converts direct current (DC) electricity to alternating current (AC) electricity.

Although most wind generators do produce AC, it's wild AC that varies in voltage and frequency depending on the wind speed and the machine's speed of rotation. It's not usable for normal applications and appliances. To get to electricity that's synchronized to the grid or usable in the home, the first step is to *rectify* that wild AC to DC. The DC is then converted — *inverted* to be technical — to AC. If going from AC to DC to AC seems roundabout, don't worry. It's the most efficient way to get here from there.

In addition to the DC-to-AC function, battery-based inverters generally convert from low voltage to high voltage. The DC voltage out of the wind generator in a battery-based system is usually 48 VDC for a normal house or 12 VDC in a small cabin or boat system. In between is 24 VDC, which is fast being left on the sidelines, an orphan voltage. The inverters bring the voltage up to 120 or 240 VAC.

Batteryless inverters also convert DC to AC, but the DC is typically already at higher voltages. The wind turbine may kick out DC voltages in excess of 100 volts (which reduces transmission losses, reducing wire size). The inverter receives that DC voltage and converts the energy into AC form, typically at 240 or 120 VAC.

Modern inverters often have several other functions built in, including the following:

- Charging batteries in a grid-tied system with battery backup or an off-grid system. Converting DC to AC is the basic job, but it's not hard to design the same machine to do the opposite: convert high-voltage AC from the grid or a generator to lower voltage DC to charge the batteries in times of low wind; I talk about batteries in detail later in this chapter.

- Integrating with and communicating with charge controllers (which I cover later in this chapter) to make smart charging a reality.

- Starting generators, either manually or automatically.

- Monitoring and data-logging loads, charging, and generation.

Examining different types of inverters

With AC electricity, the direction of the charge flow (electron flow in wires) alternates. The electrons go in one direction for a fraction of a second and then reverse direction. This happens over and over and over. One way to conceptualize this is to graph it as a wave, such as the sine wave you may remember from trig class. Different waveforms describe different levels of smoothness of this undulating wave.

If you live in the modern world, modern electricity is sine wave, and that's what your appliances desire and expect. So make it easy on yourself — get a sine wave inverter!

Don't get sucked into the marketing hype about "true sine wave." If you want to get nitpicky, find out the total harmonic distortion (THD) of the inverter's wave.

In this section, I describe the two sine wave inverters available: battery-based inverters and batteryless inverters (see Figure 15-1).

Any wind-electric system with batteries — whether off-grid or grid-tied with battery backup — must use a battery-based inverter. You make a one-time, either/or choice between a battery-based inverter (which works only with a battery-based system) and a batteryless inverter (which can only be used with a batteryless system).

Battery-based inverters

A battery-based wind-electric system gives you *storage*, the ability to use energy when the grid or wind electricity isn't available. If you want storage, you must have batteries. And if you want to use your wind electricity when the grid is not there, you need batteries and a battery-based inverter.

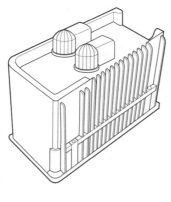

Figure 15-1:
Battery-based and batteryless inverters.

a. Battery-based   b. Batteryless

In battery-based inverters (refer to Figure 15-1a), you have three choices — square wave (ancient technology), modified square wave (old and developing-world technology), and sine wave. I recommend you choose only the sine wave, unless you're in an off-grid, low budget, developing world setting (I describe the other waves in the sidebar "The past waves of inverters").

Several manufacturers make battery-based sine wave inverters that kick out better quality electricity (the sine wave has lower "total harmonic distortion") than the utility grid. These are sophisticated devices that have many other features and high reliability. If you purchase well, you can expect 15 to 25 years or more of good service.

Batteryless inverters

Batteryless inverters that are approved for grid connection (refer to Figure 15-1b) are all sine wave — they're built to synchronize with the grid and "sell" energy to it. And truly, your selection job is usually very easy: batteryless inverters for wind-electric systems typically come with the wind generator. (I explain how to find the best wind generator for your needs in Chapter 13.)

Batteryless inverters operate only in direct connection between the wind generator and the grid. There's no way to add batteries later, nor is it possible to use the wind energy when the grid is out.

If you're looking at a batteryless grid-tied wind generator that doesn't come with an inverter, beware. These inverters need to be programmed to work with the wind generator. Some disreputable companies are selling turbines with inverters that aren't actually set up to work with the wind generator. Make sure you're dealing with a reputable supplier and common equipment.

Batteryless grid-tied systems (which include batteryless inverters) are the simplest, least costly, most efficient, and most environmentally friendly systems you can buy. The only drawback: They have no storage, so when the utility is down, you can't use the electricity from the wind generator.

Selecting an inverter with the correct voltages and wattage

Inverters must be purchased based on three basic parameters:

- ✔ **Input voltage:** For batteryless systems, the input voltage may be more than 100 volts DC. With battery-based systems, this is typically 12, 24, or most commonly, 48 volts DC.

- ✔ **Output voltage:** This is typically 120 or 240 volts AC.

- ✔ **Wattage:** This is the peak or running operating capacity of the inverter — how much energy you can use or sell at a given moment.

The past waves of inverters

In the old, old days, some people used square wave inverters. These have poor power quality, aren't terribly efficient, and can damage some appliances. You rarely find these for sale, and they don't allow you to connect and sell electricity to the utility.

In less ancient history, the market saw *modified square wave* inverters, erroneously called *modified sine wave* inverters by the marketing departments of some inverter companies. These made and still make great backup inverters for developing-world situations. Some of these inverters can use the grid for charging batteries, but none are able to sell electricity back. These are cheaper than modern sine wave inverters, and you get what you pay for.

The power quality of modified square wave inverters is lower, so if you live in a rural, developing nation with poor utility infrastructure, your appliances may be happy to use modified square wave electricity, because it may be as good as the local utility electricity. Or if you live in an off-grid situation on a low budget, you may consider a modified square wave inverter — I encourage you to seek out a high-quality unit, not the truck-stop specials out there.

For a batteryless grid-tied wind-electric system, the machine you buy comes with the appropriate inverter. The system voltage is a decision that's out of your hands.

With a batteryless grid-tied system, the manufacturer of the wind generator chooses the alternator's voltage range (the input voltage) and sells the machine with the appropriate inverter that's synchronized to the grid (the output voltage). The wattage is determined by the absolute peak power of your wind generator. Your shopping, and even your dealer's shopping, doesn't generally involve evaluating the choice of inverter.

For battery-based systems, the battery voltage you choose will be your inverter's input voltage, and the inverter's output will be the standard 120 VAC or possibly 240 VAC from one or two inverters. Your output voltage may be higher than your battery voltage if you're using an MPPT step-down controller (see the later section "Bonus features of different types of controllers"), but again, this generally comes in a package with your system.

With a battery-based system, you may also have to choose your inverter's wattage — the power capacity (that is, up to what level it'll deliver electricity). The inverter indeed needs to be able to handle the peak output of your wind generator: That is, if the machine can deliver up to 15 kilowatts (kW), you need at least a 15 kW inverter. But the inverter also needs to be able to handle your peak load — how much power your off-grid home or backup system will draw. In some cases, this will be higher than the peak output of the wind generator.

Understanding the Basics of Batteries

Most on-grid wind-electric systems installed today do not have batteries. Batteryless systems really have only one disadvantage — they provide no backup electricity during utility outages. When the utility grid is down, any system without batteries is also down, regardless of how hard the wind blows.

If you want utility outage protection, you need batteries, which are electro-chemical storage devices (see Figure 15-2). In this section, I explain the purpose of batteries in your system and describe battery types. In Chapters 16 and 19, I discuss proper sizing, maintenance, troubleshooting, and replacement. Because batteries need a lot of care and attention in real life, I give them a lot of attention in this book.

The purpose of batteries: Storing energy

If you're on-grid and want electricity for some or all of your appliances and other electrical loads when the grid fails, you need batteries. Typically, these systems don't supply all your appliances, or if they do, it isn't for very long. A common scenario is to supply critical loads (perhaps lights, communications equipment, a furnace blower, a water pump, and maybe a refrigerator) for a day or two. This results in a battery bank that's manageable for space, cost, and maintenance.

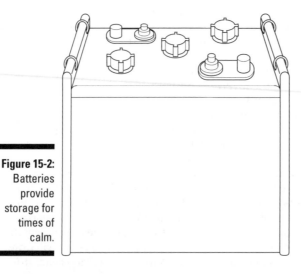

Figure 15-2:
Batteries
provide
storage for
times of
calm.

If you're off-grid, you have to have batteries. They provide a buffer to supply appliances with sufficient starting and short-term energy, usually via an inverter (I discuss inverters earlier in this chapter). They provide storage for periods of calm weather when the wind isn't putting electricity into your battery bank.

Can you get along off-grid without batteries? Well, theoretically, but only if you're happy to have electricity only when the wind blows and if you're ready to have your computer crash and your dishwasher stop every time there's a lull in the wind. That doesn't sound very practical to me, and in fact, you won't find wind power systems designed for off-grid (also called *stand-alones*) use without batteries, except for the rare system that directly pumps water or produces heat.

Looking at battery types

No batteries are made specifically for the tiny niche market of wind energy. Most non-sealed batteries used in wind-energy systems are originally made for two applications: golf carts and floor scrubbers. Knowing that you're dealing with products designed for these other industries kind of puts the little wind industry in its place. But that's the reality, and the batteries work well for wind energy.

Determining what kind of battery you're getting is often a bit hard. You don't want starting batteries, and you probably don't even want most marine/deep-cycle batteries. You want *deep-cycle* batteries, which are intended for deep discharge and recharge over hundreds of cycles. These may also be called *traction* batteries — they're used in mobile applications, such as golf carts and forklifts.

Although you can find a variety of exotic battery types, almost all systems use one of two general categories of deep-cycle batteries — flooded and sealed:

✔ **Flooded batteries:** *Flooded* means that liquid is involved and that the battery cells are open. Flooded batteries need regular replenishment of the *electrolyte* (the liquid inside the battery) because it's lost through evaporation and battery gassing. In a well-designed system, you should check electrolyte levels every few months and fill the cells with distilled water. Initially, you should check monthly, and you may gradually find that quarterly is adequate.

Never let the plates inside a battery get exposed to air. This permanently damages the battery, reducing its capacity and shortening its life. Flooded batteries also give off corrosive and explosive gases when charging, so follow the safety tips in Chapter 17.

These batteries are typically the least expensive, and they usually last longer for the dollars spent.

✔ **Sealed batteries:** Sealed batteries are just what they sound like: batteries that aren't open to the air. They can't be refilled, and no maintenance is required. Sealed batteries have one of two basic internal designs:

- One is gel, which is what it sounds like — the acid between the lead plates is in a gel form instead of a liquid.

- AGM (absorbed glass mat) batteries use fiber separators that absorb the electrolyte.

AGMs are generally more robust than gel batteries, but both require careful charging, which essentially means that your charge controller needs to be set properly. If you overcharge sealed batteries, you lose moisture, and replacing it is impossible. The disadvantages of sealed batteries are that you need to charge them more carefully, they cost more, and they don't last as long.

Weigh your options — convenience versus cost effectiveness. If you're a hands-on person or are willing to pay for regular service of your battery bank, I recommend flooded batteries. But you'll need a vented box and a maintenance schedule — at least quarterly, you or someone else needs to check battery water, connections, and do a cleaning. If this seems like too much, or you have a special circumstance (if batteries need to be on their sides or in closer proximity to people, for example), sealed batteries may be just the ticket. Be prepared for higher initial cost, earlier replacement, and charging sensitivity.

Flooded and sealed batteries are lead-acid battery technologies. Lead-acid isn't new technology; in fact, it's been around for more than 100 years. It's reliable and predictable technology, and it's cost effective. Other lead-based batteries are out there, as well as nickel-cadmium (NiCad), nickel-iron, lithium ion, and other newer technologies. In most cases, these will either be beyond your budget or present other problems, such as a different voltage range, unusual care, difficulty in recycling, and so on. I recommend that you stick with the tried and true for now.

Minimizing the drawbacks of batteries

Batteries are the only practical way to have storage, but they entail a number of costs. Compared to batteryless grid-tied systems, battery-based systems (both grid-tied with battery backup and off-grid)

✔ Are more expensive

✔ Are less environmentally friendly

✔ Are less efficient

✔ Require more maintenance

✔ Require periodic battery replacement

To lessen the drawbacks of having batteries in a system, you can take the following actions:

✔ Size your battery bank carefully so that your system is balanced. Oversizing battery banks is common, which leads to greater costs, more-difficult charging, and more maintenance. (I discuss sizing your battery bank in Chapter 16.)

✔ Include battery state-of-charge metering in your system so you know at any given time how much energy you have in your battery bank. Take the time and effort to get the meter configured properly so it gives you good information.

✔ Be aware of how you're using energy and fully charge your battery bank at least a few times a week.

✔ Perform all battery maintenance on a regular basis. (See Chapter 19 for details.)

Checking Out Charge Controllers to Protect Your Battery Bank

Connecting your wind generator to a battery bank without a device to manage the battery bank's level of charge is a recipe for disaster. So if you have a battery bank, you need a charge controller (see Figure 15-3). In this section, I explain the purpose of charge controllers and how they work, and I describe different controller types and features.

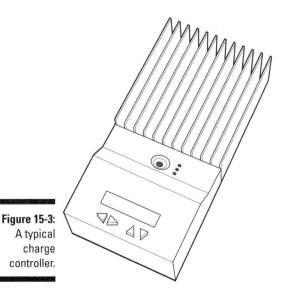

Figure 15-3: A typical charge controller.

The purpose of charge controllers: Protecting batteries from overcharging

Batteries like to be charged within a certain voltage range, and regular under- or overcharging doesn't serve them well. Here's why:

- ✔ **Undercharging:** Chronic undercharging of your battery bank leads to premature failure — and that will hit your pocketbook hard. Careful energy monitoring and use help you avoid undercharging, though some electronics (in the charge controller and inverter) can prevent radically deep discharges.

- ✔ **Overcharging:** Periodic overcharge (equalization charge) is useful for a battery bank, but doing it regularly and randomly may not be good for your battery bank, especially if you're not paying attention.

Overcharging gasses the battery bank, which means that fluid is leaving the cells. If you're not on top of watering flooded batteries, you can easily expose plates and damage the bank. Protecting against overcharging is the primary job of a charge controller. You need a charge controller to prevent excessive charging, gassing, and loss of electrolyte.

The charge control function is only in battery-based systems. This makes sense, because its primary function is to protect batteries. With battery-based grid-tied systems, you need a charge controller, but it primarily comes into play (in its charge control role) during utility outages. During normal operation, the inverter acts out the charge control role by diverting energy to the grid, not to a dump load.

Surveying series versus diversion controllers

You can find two basic configurations of charge controllers in the renewable energy world:

- ✔ **Diversion (dump load):** Diversion or dump load controllers divert energy away from the battery to achieve the charging goals (*set points* — see the nearby sidebar for details). This allows the wind generator to be fully loaded so it maintains its proper operational speed range. Most wind generators use this kind of charge controller.

- ✔ **Series:** The series controller doesn't work with most wind turbines. This controller is typically wired in series between a solar-electric array and a battery bank. It simply turns the PV array off, opening the circuit when the battery bank is fully charged.

If you were to open the circuit of a wind generator, the turbine would *freewheel*, running faster than intended. High speeds can be catastrophic for some turbines, resulting in electrical or mechanical damage. However, some specific wind generators can freewheel safely, and therefore they may use series charge controllers.

Diversion controllers need another component, called a *diversion load* or *dump load*, so you don't overcharge the battery bank. The dump load is an air- or water-heating element that's capable of absorbing all the energy the wind turbine can deliver. Air heaters are more common and may be simplest for hands-off users, because a water-heating element can't always dissipate all the heat. So you may need a backup if you choose that system. When the water gets very hot, the element may not be able to dump what it needs to without blowing the temperature/pressure (TP) relief valve on your tank.

But don't think that this is a huge amount of heat. The dump load comes into play only in off-grid systems or during utility outages in grid-tied systems. In grid-tied systems, your surplus energy is usually sold back to the utility for credit, not burned up in heat. In off-grid systems, you can minimize energy sent to your dump load by being aware enough to know when you have a surplus and finding creative ways to use it, such as cutting firewood with an electric chainsaw, pumping water, charging an electric vehicle, and the like.

In any case, with battery-based systems and most wind turbines, you need the controller/dump load combination to protect the batteries and wind generator. Note that in grid-tied systems, the inverter carries this charge control function when the grid is up, selling energy down to the battery set point whenever the grid is available (see the next section for info on set points). The controller/dump load combo gets on stage only when the grid is down.

In most cases, the charge controller is sold with the wind turbine and is specifically designed to be used with it. If your dream battery-charging machine doesn't come with a charge controller, think twice about buying that wind generator. But if you're a dedicated do-it-yourselfer, you can match up a controller with the wind generator, looking at input voltage, battery voltage, and overall capacity (wattage).

Bonus features of charge controllers

The simplest solar-electric charge controllers may have only battery-charge control (series), and it may be very rough — it's commonly called "slam-bang" control, and it's just full on or full off. But as you move into diversion controllers (which are necessary for most wind generators; see the preceding section) and into higher-end controllers in general, you find more and more other features, including the following:

- ✔ *Pulse width modulation* (PWM) charging, in which the controller sends energy from the battery bank to the dump load in little pulses of varying length; this strategy is designed to divert the necessary energy to stay steadily at the voltage set point

✔ *Maximum power point tracking* (MPPT), which gets the most out of the charging source by finding the highest power spot on the wind generator's voltage curve for each wind speed instead of simply running the wind generator at the battery voltage

✔ *Step-down conversion,* so you can use high-voltage wind turbines (and therefore smaller transmission wiring) connected to lower-voltage battery banks (12, 24, and 48 volts are typical)

✔ Metering and data-logging to document instantaneous and historical system performance

✔ *Low voltage disconnect* (LVD), which takes the DC loads offline when the batteries are dangerously low; this is not recommended for general protection of batteries from excessive discharging, but LVD stems the injury when it's severe

Other features come and go with companies and time. Perhaps most exciting is seeing charge control functionality become integrated with other system components, so that the charge controller and inverter communicate with each other and manage the energy, battery charging, and loads wisely.

On the Safe Side: Wire, Disconnects, Overcurrent Protection, and Grounding

Fortunately, the manufacturer or supplier usually handles major system component decisions, such as the inverter and charge controller. Other component decisions require you and your partners to know or find out a lot about electrical systems in general and wind-electric systems specifically.

Beyond the major components such as a wind generator, tower, inverters, batteries, and controller are many other components that you need to make a complete system. Notice that I don't call them "minor" components; although they may be relatively small in cost and size, they're not minor in importance. Four key areas are worth considering here because they're linked to safety: wire, disconnects, overcurrent protection, and grounding (see Figure 15-4).

Wire, grounding, overcurrent protection, and disconnects are complex topics that you or your installer need to pay attention to. This section is by no means an exhaustive coverage of the topics nor of the many other individual components you need to make a complete system. No book will make you an electrician, and I encourage you to partner with an electrician experienced with wind-electric systems for part or all of your project (see Chapter 12 for guidelines on partnering with experts). Don't think lightly on these topics; take them seriously, as seriously as you take your life and home.

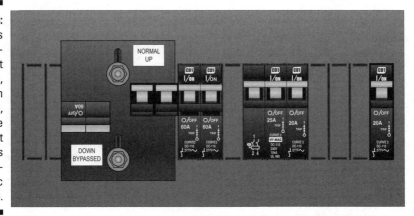

Figure 15-4:
Disconnects and over-current protection, along with grounding, are important components of any wind-electric system.

Wire: Your energy carrier

Your wind generator makes electrical energy. Your batteries store it. Your charge controller and inverter condition it. Your appliances use it. In between all these things are networks of wire. This wire needs to

- ✔ Be sized properly for safety; it needs adequate ampacity (charge-carrying ability) so it doesn't overheat, melt the wire's insulation, and burn down your house

- ✔ Be sized properly for energy efficiency; it needs not to restrict charge flow and increase voltage drop, wasting your precious wind energy

- ✔ Have appropriate insulation to protect it from the conditions it lives in

Most wiring for home electrical systems these days is copper, though you still find aluminum wiring in older homes and in specific cases. Copper is more expensive than aluminum, but it's a better conductor, which means you can use smaller wire. Copper also is less subject to corrosion. Too often, poorly installed or older aluminum wire runs (especially underground) corrode and fail.

You can size the wire by formula, but it's more often done using charts that show voltage drop for specific sizes, voltages, and wire lengths. Keeping voltage drop in the 1 to 5 percent range is typical, and this usually means that your wire is also sized safely for amperage.

Wire sizing and use is not simple. I encourage you to ask your supplier for help on sizing the main wire run from the wind generator to the controller/inverter. Partnering with an experienced electrician is also advised.

How charge controllers work: Picking up on set points

How do charge controllers work? Well, they're always looking at battery voltage (in other words, electrical pressure). They use this as a rough measure of state of charge (SOC) of the battery. Each battery also has limits of what voltage it wants to see on a regular basis. Each charge controller has *set points* (specific voltage levels) for at least *bulk* charging (fast refill of the biggest part of capacity) and *float* charging (trickle charging the battery from almost-full to full), and it may have one or more other charging stages (one common stage is *absorb*, which is somewhere between bulk and float).

When the controller sees that the battery voltage is below the appropriate set point, it does nothing, allowing the wind generator and other sources to charge the battery bank. When the controller sees that the set point has been reached — perhaps it's 57.6 V — it prevents energy from reaching the battery to keep the voltage at the set point.

Battery charge set points are different for every battery technology and each specific product. Talk to your battery manufacturer or supplier to make sure that you have the right charging set points for your specific battery.

Disconnects: Shutting it down

A *disconnect* is a fancy name for a switch. It's a means to electrically disconnect one component or set of components from another. In your normal interaction with switches, they're primarily for convenience. You want to be able to switch your lights and appliances on and off when you want to.

Disconnects in wind-electric systems are for convenience and safety. You want the ability to isolate each part of the system when maintaining, troubleshooting, or repairing the system. Most places are subject to some form or other of electrical code, and these codes generally require a disconnect that can isolate every source of energy from every user of energy.

Safety disconnects typically come in the form of *circuit breakers,* commonly just called *breakers.* Purchase them with appropriate voltage and amperage ratings in mind.

A disconnected wind turbine produces two to five times its normal voltage, damages electronics, and may be a hazard. Take care when disconnecting the battery of a wind-power system, and first make sure that the wind turbine has been disabled via mechanical or electrical braking. Often you can do this by short-circuiting the turbine, but in some cases, it's better to disconnect it at source where the wires enter the building. Check with the manufacturer to find out how best to isolate the turbine while working on the system.

Overcurrent protection: Automatically breaking the circuit

All electrical circuits should have overcurrent protection. Overcurrent protection has a separate function from disconnects, though often the two functions are in one package: circuit breakers. *Overcurrent* protection is just what it sounds like: protection from high current. If an electrical component fails or for whatever reason you have high amperage in a circuit, you want to protect the components and wire from damage.

A breaker has a certain amperage rating, and if the flow of charges in the circuit exceeds that rating for the specified time, the breaker trips. (Common language says that you've *blown, popped,* or *tripped* a breaker, though some of those terms are more suited to another form of overcurrent protection, fuses, which have a small wire inside that melts if the amperage exceeds their rating.) Because breakers are resettable, they're designed into systems much more often than fuses, which are one-time-use devices. Whichever form of overcurrent protection you use, it needs to be sized so that it won't operate during normal use of the circuit but will trip before the wires get overloaded.

Grounding: Offering excess charge a way out

If you want to really get confused, delve into grounding, one of the most complex and contested subjects in electricity. Perhaps one reason it's so tough to grasp is that grounding actually has three different purposes:

✔ To prevent shock and electrocution to end users

✔ Combined with breakers, to let users know that a problem is in the wiring (a *ground fault*)

✔ To reduce damage from lightning

The basic strategy is to run a dedicated wire that connects all system components and that often also connects to metal ground rods driven into the ground. This allows any fault currents to head to ground instead of going through you or the equipment. On its way to ground, the current breaks the circuit breaker, alerting you or your electrician to the problem.

You want your tower to be grounded to protect it from lightning. Depending on the rules you're subject to and whether you live in a high-lightning area, your grounding system may be any of the following:

- ✔ A rod at each anchor and at the tower base, wired to the guy wires and tower

- ✔ The same as the preceding bullet but with all rods tied together, perhaps with a large grid of copper wire surrounding the tower

Other lightning protection can be complex and expensive. One manufacturer of a wind generator and inverter combo that costs about $8,000 charges about $2,500 for a lightning-protection package if you want the product warranty to include lightning damage.

Want to know what to look for in a grounding system? Look for a grounding expert to guide you. This is an important topic but not a simple one. Seasoned experts in the renewable energy industry have long discussions about these issues, some of which I can't even grasp the details of. Make sure you're on solid ground by finding local experts who know the intricacies of the topic and the local codes and practices.

Odds and Ends

The following components may escape your notice as you drool over the smooth wind generator blades, shiny inverter, and high-tech charge controller, but they may be crucial parts of your system:

- ✔ **Voltage clamp:** This component lives between the wind generator and the charge controller or inverter, and it limits the voltage output to within the limits of the equipment that follows it. This is typically supplied with the wind generator if needed.

- ✔ **Utility meters and disconnects:** If your system is grid-tied, your local utility will require these components for their purposes and perhaps for incentive payment. They're specific to each utility and state, so do your homework.

- ✔ **Distribution panel (circuit breaker box):** This will already be built into an existing home, and your system will tie into it. If you have a generator backup system, you may also have a subpanel for the circuits that you've chosen to back up.

Chapter 16

Tying Everything Together: System Sizing and Design

In This Chapter

▶ Sizing the system correctly

▶ Sourcing your own system

▶ Collecting and assembling components

*P*ulling all the threads of your research and education together into one whole piece may stretch you, but it's also satisfying in the end. Before you take the steps in this chapter, I suggest that you review your motivations and goals (see Chapter 2) and get a firm grip on whether you want and/or need a batteryless grid-tie system, a grid-tie system with battery backup, or an off-grid system (see Chapter 9). In this chapter, I describe how to size and gather the components you need.

With this work done, you're not out of the woods, but you're prepared to get your — or someone else's — hands dirty with the installation (see Chapters 17 and 18). In my teaching, I speak to contractors who have great building skills but are skeptical of the theory, design, and planning lessons, desiring only technical, equipment, or installation details. But planning and design are the most important parts of the process; without them, construction skills aren't worth much.

Good system design gives you the possibility of good installation and good performance. Without good system design, you'll be lucky to achieve your goals, and you may achieve a waste of money, a disaster, or both.

Sizing Important System Components

Correctly sizing the major parts of a wind-electric system (the generator and the tower) is a multi-step process with several branches. Other gear may need sizing decisions, but most often, charge controller and inverter come as a package with the wind generator, and transmission wiring can be sized with advice from the manufacturer. The basics, which I describe in the following sections, look like this:

- ✔ Size the generator:
    - Analyze your load and energy efficiency.
    - Assess your wind resource.
- ✔ Size the tower:
    - Choose your tower site.
    - Choose your tower height.
- ✔ Size your battery bank (if necessary).

This process isn't always linear. For one, your load analysis and efficiency assessment affect each other. You may come up with a preliminary load number but then see some opportunities for efficiency improvements that reduce your load dramatically. Wind resource, site selection, and tower height are all intertwined as well. And increasing your wind resource through tower site or height choice decreases the size of wind generator and balance of systems (BOS) components you need. So although you need to take this process one step at a time, you also want to loop back through the process as your decisions make changes in the overall picture.

Your goals and limitations also affect the process. For instance, you may have a specific budget that keeps you from sizing a system that covers your whole load. Or you may have legal restrictions that affect the design (Chapter 2 discusses restrictions you may encounter). Making a wind-electricity system work is no easy task, so it's no surprise that the sizing process has some complexities as well. Take the time necessary to do a good job at this stage. You won't regret it.

Using load, efficiency measures, and wind resource info to size your generator

If you're trying to dig a hole on a certain budget, you have to make some choices. Do you hire a backhoe, buy a shovel, or find a laborer? You look at each option's cost and effectiveness and choose wisely.

Wind-generator sizing is a similar process of matching the job to be done to the resources available. But it's actually a simpler process, if — and this is a big *if* — you do the homework to accurately estimate your energy load and your wind resource. Wind generators have a certain collection area (the *swept area*) and can produce a certain amount of energy in a given wind resource. This part isn't rocket science. Get a handle on your load and wind resource, and you can quite easily see your wind generator options.

Load analysis

Electrical energy is measured in kilowatt-hours (kWh), and that's the measure you need to understand and use when you analyze your *load* (your electrical energy use). I use kilowatt-hours per day as a basic measure of household energy use. You can talk in weeks, months, or years, but looking at one day seems more direct to me. If this measure is new to you or you just want a refresher, head to Chapter 4.

If you don't know how many kilowatt-hours of electricity you use per day, drop your wind energy dreams for the moment and do this work! Even if you think you're just going to mess around with wind or don't expect to make all your electricity with wind, you need to know what you're actually using and what a system can contribute. This legwork is your baseline for comparison. You need to be able to have a clear, honest discussion with your seller about how much energy a system produces and how that matches up to your needs.

As you prepare to analyze your load, keep the following points in mind:

- ✔ A typically wasteful North American home that doesn't heat with electricity uses between 25 and 30 kWh per day.

- ✔ Average daily energy use is just that — an average. Some days have a much higher usage and some have a much lower usage. Your daily use is also different in different seasons.

- ✔ Your wind resource also varies from day to day and season to season. So if you're off-grid or have a *monthly zeroing* of your utility's net metering policy (that is, your utility cancels out any credit you have from surplus energy on a monthly basis — not a very good system for you), you want to do your design work looking at the seasons individually.

In looking at clients' homes, I've seen modern, fully featured homes that use only 8 or 10 kWh per day. I've seen other, similar-looking homes that use 40 kWh per day. Though size plays into these differences, the larger factor is the energy appetite of the occupants.

On-grid load analysis can be as simple as looking at your utility bills for the past year. For off-grid homes or new construction, it's more complicated and may mean doing a load-by-load measurement, calculation, or estimate of each load; see Chapter 6 for details.

Using less energy

After you have a kilowatt-hour per day figure for your home, you're not ready to break out those wind-generator brochures just yet. Revisit your goals — if they include saving money or saving the planet, you want to take a hard look at options for energy efficiency. (If you only want to see things spin, I suggest pinwheels and other rotary yard art.)

Before you decide which size generator you need, look to see whether you can reduce your energy usage through *conservation* (turning off the lights when you're not using them) and *efficiency* (using less energy to get the same amount of light). Finding ways to be more efficient can help cut back your energy load and reduce your required generator size.

If you read this book and never do anything about wind energy, I hope you still work on energy efficiency in your home. It's the most fruitful field for saving dollars and the environment. In my opinion, putting in a wind generator without evaluating your efficiency first is a big waste of money, time, and resources. Chapter 7 has plenty of tips for increasing your home's energy efficiency.

The potential of your wind resource

The wind resource is one of the most misunderstood factors of any energy equation I know. Most people have a pretty good grasp of how much energy is in a gallon of gas, at least to the extent that they know it gets them from here to Grandma's 35 miles away in their sedan.

Wind energy is much tougher to understand or even get an intuitive sense about. It's invisible, elusive, and variable in a way different from what you're probably used to. Its energy varies with the cube of the wind speed, meaning that double the wind speed gets you eight times the energy potential, not two. (Chapter 5 has more on this concept.) Getting your system to reach strong, smooth winds is important; jumping up a few miles per hour can mean doubling your energy.

You also need a reasonably clear idea of what your wind resource is. Saying that it's 10 or 12 mph average is much too rough (the difference between these two figures is about 70 percent in energy availability). Ideally, you should get a long-term measurement or estimate that has an accuracy of 1 mph, if not ½ mph. Be conservative if you're estimating or analyzing figures to get that number. Chapter 8 has more on understanding average and instantaneous wind speeds and on siting and determining your site's average wind speed.

Using your numbers to determine a wind generator size

After you know your load and wind resource (see "Load analysis" and "The potential of your wind resource" earlier in this chapter), the rest of your work is easy. At this point, you need to rely on projections of production for a variety of wind turbines. I recommend looking at all possible sources. Sources for these numbers may include the following:

- ✔ Manufacturers' energy curves (see Chapter 5) — but also find field data from unbiased sources to confirm or refute the manufacturer's estimates, which are frequently misleading

- ✔ Manufacturers' power curves put through a wind calculator spreadsheet (the power curves alone do not have the information you need, as I explain in Chapter 5)

- ✔ Published data from public sources, such as the National Renewable Energy Laboratory (NREL) or Appalachian State University (ASU), or other independent testers of wind generators

✔ Published or unpublished data from private sources, such as individuals, businesses, and organizations that use the wind generators you're considering

One source that pulls together a great deal of information is an article wind-energy expert Mick Sagrillo and I co-authored for *Home Power* magazine. "How to Buy a Wind Generator" features a four-page spread that gives detailed specs on the most viable wind generators (22 in all) for homes in North America (you can find the article at www.homepower.com). In this article, we list predicted annual energy output in kilowatt-hours for 8, 9, 10, 11, 12, 13, and 14 mph average wind speeds. With this data gathered in tabular form, you can easily crunch some numbers and see what your options are. All you need to do is multiply your daily load by 365 for a yearly energy usage total and then find your average wind speed in the table to discover the generator options at your load level.

For example, suppose you determine that the average daily energy use in your home is 22 kWh and your average yearly usage is 8,030 kWh per year (or 8 megawatt-hours [MWh]). If you've determined that your average wind speed is 9 mph at *hub height* (the height at the center of the wind generator's rotor) on your site, you can scan across the table in this article and find the possibilities.

In this example, the table shows three turbines that project production between 10 and 11 MWh per year in a 9 mph average wind speed. Below this, the next turbine projects only 5.6 MWh per year (significantly below the 8 MWh you use), so I suggest confining your choices to the three. At that point, you can begin to look at other characteristics before making your final choice. (Chapter 13 describes various generator traits you should consider.)

Bear in mind that this article is just one source of numbers. It's a handy source, but don't rely on any one source alone. Check out the information sources earlier in this section for more complete info.

If you don't have good, independent confirmation of energy production on turbines you're considering, you may need to use other methods. *Swept area* — the area of the circle the rotor makes as it spins — correlates relatively well with production, so using a simple formula, such as the one I introduce in the Cheat Sheet, may come in handy. I encourage you to use as many methods as possible to come up with the best numbers you can. Don't stake your sizing and buying decisions on only manufacturers' estimates!

Sometimes this turbine-sizing exercise has to circle back. Here are a few circumstances that may require some recalculation:

✔ After you calculate your load and wind resource, you may discover that the turbine you need is out of your price range. At this point, you need to first revisit energy efficiency (see Chapter 7) to see whether you can reduce your load and therefore your cost.

✔ You may also determine that you can't afford to purchase a wind-electric system that provides all your electricity or year-round electricity off grid. That's okay. Making 20, 40, 60, or 80 percent of your electricity with the wind may serve your goals just fine; you can use the utility grid, solar electricity or hydroelectricity (see Chapter 11), or a fuel-fired generator to provide the rest of your electricity, depending on your situation and goals.

This situation is also a good opportunity to reconsider your proposed tower height (see the next section), which affects your average wind speed. You may find that increasing your tower height by 20 percent increases your production enough that the same wind turbine that would've done 70 percent of your load can now do it all.

Keep walking these numbers around until you find the balance that fits your load, your resource, and your budget. And don't forget to plan for the future. Although you may not be able to afford the wind generator of your dreams today, perhaps you will in two, five, or ten years (although you should reevaluate your needs at that point — they may change). Put in a strong enough tower to carry the biggest wind generator you ever expect to own (flip to Chapter 14 for more on tower selection). And make it as tall as you'll need for it to stay 50 feet or more above the trees for the rest of your life.

Using your tower site and wind shear to determine your tower height

Tower height is a crucial decision. This decision is where people make the biggest and most irrevocable mistakes. After you put concrete in the ground and engineer the specifics of your tower design, you're pretty much stuck. You're not going to upgrade your tower to make it taller — you'll need a whole new tower, from the foundation up.

At the level of the highest *obstruction* (tree, house, swing set) on your site, you have a certain average wind speed. As you rise above the obstructions, the wind speed increases. How much it increases depends on the *wind shear,* which describes how quickly the wind speed increases as you go higher (see Chapter 14). Understanding your site and wind shear can help you make a wise decision about tower height. Taller towers mean higher wind speeds, which mean higher energy. Going for a taller tower than you think you need is the simplest, most cost-effective move you can make in your system design.

The right site for your tower

As I mention in Chapter 8, choosing where on your property to install your tower involves several factors:

✔ The highest point on the property likely has the most wind and will allow you to get at the best wind resource.

Being realistic about higher average wind speeds

You may wonder why the table in the article "How to Buy a Wind Generator," which I refer to in this chapter, doesn't give energy production at higher average wind speeds than 14 mph. Even 14 mph is beyond what most people are lucky enough to live with, as I note in Chapter 5. It's the low end of wind farm resources, and the very high end of home-scale resources.

After seeing one disreputable company publishing an energy curve that goes up to a 30 mph average, I did a little research on what manufacturers are advertising in terms of average wind speed. I found that the best companies don't publish numbers above 14 mph. I asked the sales director of one company why they don't give higher numbers. Here's his response:

"Annual average wind speeds over 14 mph are few and far between in the real world. Newbies all think that *their* site is uber-windy and tend to look towards the high end of the published data when they 'imagine' their performance. We cut their dream back to a level that remains within the realm of possibility. . . ."

Now this is the kind of sales pitch I like to hear! Most home-scale wind sites are in the 8 to 12 mph range, with 12 being very healthy. Being realistic about this saves customers, installers, and manufacturers from disappointment and backlash later.

- Measuring and mapping the heights of nearby obstructions can help you determine a more realistic ideal height. Obstructions between the site and the prevailing wind have the largest impact on the wind available to the turbine, but downwind obstructions also affect performance. Know your wind direction distribution through a graphic called a wind rose.

- The distances between potential tower sites and your home or utility connection affect your choice of site. You may have to go with your second or third site choice if the best site is far enough away that the cost of wiring the electricity to where you need the energy is prohibitive.

- Tower style (see Chapter 14) also affects potential tower locations: tilt-ups take large flat areas, fixed guyed towers take less open space, and freestanding towers have very little footprint.

After you've identified the best potential sites, map your property and identify all potential sites, with the preceding information noted, to make clear comparisons as you proceed with your choosing process.

Crunching some numbers to come up with a tower height

With the data you collect about your ideal tower site(s), you're ready to do the math and determine your ideal tower height. Here are the first few steps:

1. Determine the average wind speed on your chosen site.

To make any kind of intelligent system design, you must have an estimate of the average wind speed on your chosen tower site. But rarely do you ever have long-term data for exactly your tower height on your site. More likely, you have some data on your site or at a neighbor's, or you're using wind map data, or you're approximating based on data from a few miles away. Get the best data you can, preferably from several sources (see Chapter 8).

2. **Shear that wind speed data to match your proposed tower heights.**

   Most of your wind data probably won't be at your proposed tower heights. So you may need to shear, or adjust, the data up or down. Essentially, *wind shear* means that if you're in rough terrain, the wind speed increases very quickly as you go higher; if you're in smooth terrain, it increases more slowly as you go higher.

   If you're a glutton for punishment, you can seek out the formulas and crunch all the numbers yourself. But if you want to save your brain and your calculator-punching fingers, the Danish Wind Energy Association (DWEA) has made the wind shear job easy. Head to www.windpower. org/en/tour/wres/calculat.htm for the DWEA's calculator. You input your average wind speed at a certain height, and the calculator tells you what the average wind speed will be at several other heights. It also tells you what the speed will be in different shear conditions, which are identified through the following roughness classes:

   - **Class 0:** Open water

   - **Class 0.5:** Open terrain with smooth surface, such as airport runways

   - **Class 1:** Open agricultural areas without fences and hedgerows

   - **Class 2:** Agricultural land with some hedgerows and scattered buildings

   - **Class 3:** Villages, towns, agricultural land with many hedgerows and buildings, and forested land

   - **Class 4:** Cities with tall buildings

   This complex calculator saves *a lot* of work and gives you a better picture of what your wind resource is. But you can't start with no or bad data and expect this calculator to bail you out. You have to do the legwork to give the calculator the best data possible.

3. **Look at the manufacturer's information for your generator to determine how much electricity the generator creates at these heights.**

4. **Price towers at each height.**

5. **Compare total costs for purchase, installation, maintenance, and the like.**

6. **Compare total energy produced at each height with your energy need.**

   That's where the load analysis from earlier in this chapter comes in handy — see the section "Load analysis."

7. **Buy the tallest tower you can afford.**

Suppose you bought off-grid property in remote northern Minnesota and you're trying to size a wind-electric system. You have few or no neighbors, and no local wind data that you know of. You're going to have to rely primarily on the wind mapping. You check with two sources and find the closest data to your expected tower height is at 50 meters (164 feet). The data you have shows that a 9.4 mph average wind speed at 50 meters.

Your preliminary generator sizing exercise leads you to a 22-foot diameter machine (see the earlier section "Using load, efficiency measures, and wind resource info to size your generator" for info on sizing). Your land is flat, as is the surrounding countryside, and mature tree height is about 21 meters (70 feet). If you use the 30/500 rule — in which you want the hub of your turbine to be at least 30 feet above any obstructions within 500 feet — your bare minimum tower hub height is 120 feet (30 feet above your 70-foot trees plus 11-foot blade radius). But I recommend running numbers at various heights, and using the existing tree height as the "ground." Suppose that you're considering towers at 110, 120, 130, 140, 150, and 160 feet.

You also need to determine what your wind shear is (pulled from the roughness class numbers I mention in Step 2). Again, be conservative. If you're not sure whether it's one or the other, choose the higher shear. For this example, assume that the shear is 3 rather than 2.

Now take the data you have and shear it to these heights. Your best bet is to crunch the numbers for each set of wind data you have, but if you only do one, pick the best data you have nearest your proposed tower height. In this case, it's the 50-meter wind map data because it's closest to the height of your tower. Shearing it to the proposed tower heights and converting roughly from meters to feet gives you your estimated wind speed at that wind shear. With a 110-foot (34-meter) tower, your estimated wind speed is 8.0 mph. Here are the figures for the other towers:

- **120 feet (37 meters):** 8.4 mph
- **130 feet (40 meters):** 8.7 mph
- **140 feet (43 meters):** 8.9 mph
- **150 feet (46 meters):** 9.1 mph
- **160 feet (49 meters):** 9.4 mph

Now you can use these numbers to choose a tower height. Try not to factor in your fear of heights (just get your brother-in-law to go up there). Seriously, just look at the numbers, using a specific 24-foot wind generator. Here's how your possible tower heights relate to the manufacturer's energy performance predictions at these wind speeds:

Tower Height	Average Wind Speed	Energy per Year (kWh)	Energy Increase over the 110-foot Tower
110 feet	8.0 mph	8,256 kWh	
120 feet	8.4 mph	9,632 kWh	17%
130 feet	8.7 mph	10,664 kWh	29%
140 feet	8.9 mph	11,352 kWh	37%
150 feet	9.1 mph	12,040 kWh	46%
160 feet	9.4 mph	13,072 kWh	58%

Now I'm not going to walk you all the way through this example, but I hope you get the idea. The remaining steps involve pricing towers at each height and comparing costs and energy output. Chapter 10 tells you how to use pricing and energy info to calculate cost per kilowatt-hours so you can see how various turbines and tower heights compare.

You can't grow a tower. Your initial height selection is your one chance to increase the performance of your dream wind generator, so please, please make it thoughtfully. Note that using the shear calculator in rough terrain may not be as simple as it looks. Certainly you can't shear from true ground level in those cases — there's little or no wind below the tree line. So you may need to think of the canopy top as the ground and go from there. If you're at all in doubt, be conservative — taller tower and larger rotor.

Sizing your battery bank (if you have one)

If your system will have batteries, how big a battery bank do you need? This is not an easy question. It depends on a number of factors, some of which are quantifiable and others of which are squirrelier. I start by distinguishing two different system types, because the decision-making process is very different for off-grid systems than it is for on-grid.

Sizing batteries for off-grid systems

As you find out in Chapter 9, off-grid systems must make all their energy one way or another, with no grid to fill in the holes. These systems almost always have batteries, and they usually have more than one energy source. You never find an off-grid home system that's solely wind powered. At a minimum, you find a solar-electric (PV) system in the mix and probably a backup fuel-fired generator.

You may think that the battery sizing exercise is to discover how many days you want to go without wind (and/or sun), but if a backup generator is in the system, the actual question is how many days or hours you want to go before you have to start the generator. Limiting the number of hours you use the generator is the best way to have it serve you well for years. So striking a balance between generator runtime, wind generator size and production, and battery bank size is a puzzle worth pondering.

Batteries are not a source of energy. They only store energy; they don't make it. Moreover, they use up energy in the storage process, because they're not 100 percent efficient. In fact, they're 80 to 85 percent efficient when new, so if you give them 10 kWh, you get about 8 kWh back out. Making your battery bank larger "because you want more energy" is actually working against your goals, because the larger the battery bank, the more losses you have.

Not only do smaller battery banks cost less upfront, but they also waste less energy over time, take less maintenance, and cost less for the inevitable replacement every five to ten years. Of course, there are extremes in any design process. If you make your battery bank too small, you'll be wasting energy off-grid whenever it blows too much and you can't store the energy.

Make a careful sizing choice upfront — it's best not to change your mind later and upgrade the size of your battery bank. Although you can do it in the first year or so without harm, as batteries age, they lose capacity, and adding new batteries to old is like harnessing a race horse to a nag — in the end, you'll get only about as much capacity as the older batteries.

Sizing your battery bank involves these basic steps:

1. **Determine your total daily energy load in kilowatt-hours (see Chapter 6 for details).**

2. **Decide how many days you want to go with no charging sources (wind, sun, or backup generator).**

   Perhaps you often get stretches of two to three days with little renewable input.

3. **Multiply the daily load by the number of storage days you want to arrive at total kilowatt-hours of storage you need.**

   If your loads use about 8 kWh per day, a usable battery capacity of about 24 kWh would give you three days of runtime without wind or firing up the generator.

4. **Double the total kilowatt-hour storage you need to allow for a 50 percent depth of discharge (DOD) for the battery.**

   To maintain the recommended maximum 50 percent DOD, you'll need 48 kWh of battery capacity total.

5. **Divide kilowatt-hour capacity by the nominal battery voltage to determine the amp-hours of battery capacity required.**

   Here are your possible choices for nominal battery voltage:

   - **48 volts:** If you're talking about a typical American home, you want a 48-volt battery bank. That's the standard these days. Going higher voltage runs into code issues and nonstandard equipment. Going lower runs into high losses and/or expensive wire.

   - **12 volts:** If you have a small cabin or boat, a 12-volt system may be appropriate.

   - **24 volts:** You may consider landing in between at a 24-volt system, and one popular small wind generator is available only in 24 volts. However, 24 volts is nonstandard these days. As someone who lives with such a system, in most cases I'd urge you to choose between 12 VDC (very small systems) and 48 VDC (most systems).

   You're working with a whole-house system, so you choose 48 volts. Dividing 48,000 by 48, you come up with about 1,000 amp-hours.

6. **Specify specific batteries to make up a battery bank.**

   You'd need 24 of the common 6-volt L-16 batteries, which are 350 amp-hours at 6-volts, to get 1,050 amp-hours at 48-volts, a bit more than you designed for.

Sizing batteries for grid-tie systems with battery backup

If you're on-grid, the situation changes radically from the one I describe in the preceding section, as does the purpose of the battery bank. Now instead of dealing with periods of low wind, you're worrying about periods of utility outage. These systems often do not have a generator, so the battery bank must take on loads during an outage, with only wind assistance. Fortunately, wind matches up with outages, because many outages are caused by high-wind events.

Most renewable energy system designers would recommend isolating critical loads in this situation. Backing up all the loads would be quite expensive. Having the enormous battery bank necessary to do this also doesn't make a lot of sense. So specific vital loads are usually separated into a dedicated subpanel, and these are the only loads that are usable during an outage. These loads may include your refrigerator, some lights, your furnace blower, key electronics, and perhaps communications equipment, such as phone or radio, depending on your needs and priorities.

The sizing process is almost identical to sizing for off-grid, which I outline in the preceding section. Here are the two differences:

   ✔ Instead of the complete load, it's typical to backup only selected loads, due to the high cost of batteries for the limited application.

✔ You need to decide how long a typical or worst case outage is — or whatever size you want to plan for.

Remember that these batteries will be sitting doing nothing for you for almost the whole year. Actually, they'll be doing less than nothing — they'll be using some of your precious wind energy to maintain a full charge, ready for an outage.

Sourcing and Arranging System Components Yourself

If you're hiring pros to design and install your system (see Chapter 12), they'll source and arrange system components for you. But if you plan to do your own design, you have your work cut out for you. (A hybrid option is to source the components but have a pro install them; this option isn't great, though, because any pro worth working with will want to be involved in the design and specification process.)

I hope I've said it enough times in this book: Putting together your own system can be a big job! The biggest job is educating yourself about the design and installation principles so you can make smart buying decisions. No matter how much the slick marketing campaigns want to convince you otherwise, this process isn't an easy endeavor if you want long-term, reliable wind energy. A wind generator isn't an appliance. It's more like a heating or plumbing system — designing and installing it well takes expertise and experience.

This isn't to say you can't source and design a wind-electric system by yourself. But most average homeowners just can't do it cheaply, easily, and successfully without a lot of education and excellent help. You must take the time to master what's essentially a whole new field with multiple facets:

✔ Laws/bureaucracy (see Chapter 2)

✔ Wind-energy principles (Chapter 5)

✔ Topography and siting (Chapter 8)

✔ Wind generator mechanics (Chapter 13)

✔ Foundation excavation, steel, and concrete (Chapter 14)

✔ Tower mechanics (Chapter 14)

✔ Electrical and electronics (Chapter 15)

✔ Tower climbing or tilting and associated safety (Chapter 17)

You reduce your cash out-of-pocket costs by 20 to 30 percent when you source your own system, but you also increase your time requirement and responsibility dramatically. Take a clear look at the benefits and responsibilities, and make a wise choice. If you're up for the job, go for it, but be sure to get the training and heed the warnings I describe in Chapter 12. Just realize that it's not a quick, simple, or cheap job.

Although I hope I've been clear about level of cost and responsibility, I also want to be clear about the benefits. If you're willing to do it right, designing and installing your own system can be extremely satisfying. Having lived with wind energy for 25 years, I know the satisfaction of seeing the wind make my electricity. It's even more satisfying when you've started from scratch, learned about the technology, and implemented it yourself.

In this section, I explain how to find qualified suppliers, choose system components, and put them together in a basic design.

Finding and qualifying suppliers

If you choose to do your own system, you need to find suppliers. This stuff isn't available at the local big-box store. In a small industry such as wind energy, you're dealing with specialty suppliers that usually aren't local and aren't always easy to find in general. Plus, you don't just want to trust your investment to the first supplier you find; you need to make sure suppliers are reputable. In this section, I describe what you need to seek out in suppliers.

Ideally, you can use a wind installation company in your area that's willing to sell equipment to you and give you some support. If you're looking for a local dealer, *Home Power* magazine has a paid advertising section with renewable energy installers and a free listing service on the Web site (www.homepower. com). The North American Board of Certified Energy Practitioners (NABCEP; www.nabcep.org) has certified installers, some of whom do wind. And your state's solar energy society may be able to help you find experienced dealer/installers. More likely, though, you have to buy from a manufacturer or distributor in another state. If that's your best option, I recommend going straight to the wind generator manufacturer. The manufacturer can usually supply you more or less with a complete system, including wind generator, tower, and electronics.

Looking for long track records

In an era of heightened interest in renewable energy, too many new companies just don't have the experience to know what they're talking about, so look for a company with a long track record. What's a realistic track record? Ten-plus years would be great for a locally based company, but you'll have a hard time finding that in many places. If you can't find anybody locally with that kind of experience, three to five years of wind-electric system design experience, with a dozen systems sold and operating, is a reasonable history. For a manufacturer or national distributor, look for ten years in the field and 100-plus machines operating.

Verify any company's track record by asking for references and actually talking to past customers. If possible, I strongly recommend that you physically visit installations, see the machines running in a variety of wind conditions, and talk with the owners.

Scrutinizing warranties offered

Find out exactly what the terms and conditions of the warranty are. Read the fine print and ask the hard questions to satisfy yourself that you know exactly what warranties the supplier is offering. A good supplier should offer a minimum of a five-year warranty on a turbine.

Determine whether you or the dealer takes responsibility for enforcing warranties and how willing your supplier will be to help you. Ask for names of past clients, especially those who have had warranty issues, so you can interview them about their experiences with the company.

Understanding services offered

Even if you're going to do the service yourself, understanding the skills and offerings of your supplier will help you gauge the supplier's expertise and experience. What kind of system service does the supplier offer you? Service ranges from nonexistent to full. Brand-new bootstrap companies may be selling equipment that they've never installed and may be very ignorant about its capability, reliability, and installation, which means you're probably on your own when you need to service the system. On the other hand, companies that have been at it for decades and support and service everything they sell know every nut and bolt of the equipment because they've been manufacturing, installing, and supporting it for years.

Assembling your components into a successful design

If you decide to be your own contractor, you need to decide how much of the design and component selection and collection you want to do yourself. Decide how your skills and experience match up with the task and then make a plan. This section walks you through this process.

Collecting your components

As you get ready to pick out your system components, you typically find yourself somewhere on the continuum between buying a complete package and specifying every individual component yourself. The less experience you have, the further you should move toward the package end of that spectrum:

✔ **Package:** The safe option is to look for a package that covers some or most of the components you need. You probably won't be able to buy all of them at a one-stop shop, but you may be able to do it in a few big chunks. (Chapters 13, 14, and 15 can help you figure out the traits you need in each component.) Basic packages may include

- Wind generator and electronics

- Wind generator, electronics, and tower

- All system components

Finding a wind energy installer or distributor who can survey your site and your energy needs and put together a package specifically for you is the safest path. You may spend a bit more, but you save in headaches later by going with seasoned veterans who can help you avoid the mistakes they've already made.

✔ **Piecemeal:** If you count all the components of a complete wind-electric system, including each small electrical part, you end up with a list of perhaps a few hundred items. Picking all the correct bits and pieces as a newbie is a big challenge.

To make all these choices correctly, you have to fully understand all the big-picture design decisions, electrical design and code issues, and many mechanical parameters. The chances of the average homeowner becoming an expert in each field are pretty slim. Unless you've spent a lot of time educating yourself, I recommend that you have someone with direct experience on your team to help you select components.

Digging into design

In most cases, you're going to need help at the design stage, unless you choose to spend a lot of time educating yourself. At a bare minimum, I recommend that you buy your equipment from a supplier who can supply you with more than equipment — you'll also want advice, expertise, and perspective on your choices.

Choosing your system type (see Chapter 9) will help narrow your design a bit. Understanding your wind resource (see Chapter 8) and your energy load (see Chapter 6) will help you focus on the wind generator size (see Chapter 13) and the tower style and height (see Chapter 14) that you need. Perhaps the most complex design decisions will involve the balance of systems (BOS) components (see Chapter 15). If you buy a package, some or most of the brainwork will be taken out of the picture for you. If you buy piecemeal, prepare for cerebrum stretching and time for making decisions. At the same time, you'll need to deal with permits, engineering, codes, and such.

This is a big job! Don't take it lightly. If you design poorly and install well, you will probably end up with a poorly performing system. Design well, install well, and reap the energy in the wind well!

Part IV
Installing and Operating Your System

The 5th Wave   By Rich Tennant

"The good news is, the wind turbine is working great, the bad news is, it blew all the solar panels off the roof."

In this part . . .

System installation is the topic at hand in this part. First and foremost, I look at safety precautions in Chapter 17. Chapter 18 walks through the installation process and lays out basic procedures. And Chapter 19 discusses living with, monitoring, and maintaining wind-electric systems.

Chapter 17

Safety First!

● ●

In This Chapter

▶ Knowing the potential hazards associated with a wind-electric system

▶ Mastering a safe tower climb

▶ Taking care on the ground

▶ Working properly with cranes and tilt-up towers

▶ Relying on good communication

● ●

*I*f you don't focus on safety, you may not live to enjoy your wind-electric system. Wind-electric systems are without a doubt the most dangerous renewable energy systems; they have a variety of serious hazards, some of which can lead to death or major injury.

Understanding the hazards upfront puts you in the best position to avoid them. In this chapter, I give you a guided tour of the hazards you're up against, from the old standby gravity to mechanical and electrical hazards, the safety concerns associated with climbing towers and serving as a ground crew, and more. And each hazard has tools, supplies, and procedures that can help you stay safe as well as comfortable as you work on your wind-electric system.

Safety is partly gear and partly training. After you have these pieces, the biggest part is *attitude*. Take the time to do the job safely, and you'll live to do more such jobs in the future and share your ability and knowledge with others.

Being Aware of the Hazards

The comic-strip character Pogo once said, "We have met the enemy and he is us," which sums up the bottom line on safety for wind-electric systems. No great bogeymen are out to get tower climbers. Most if not all accidents are due to climbers' own ignorance, carelessness, or disregard for having and using proper equipment.

Understanding the dangers is the first step toward avoiding them. Preparing in advance for the following types of major hazards helps you work without surprises that can injure you or others.

Gravity

Perhaps the biggest hazard with wind electricity is gravity, an unforgiving force that pulls people and objects towards the earth. Gravity is in direct opposition to your goal of getting the wind turbine high above the earth. (Check out Chapter 14 for more on tower height.) It can be a useful tool — imagine trying to work without the predictable knowledge that tools and gear will stay in your hand or on the ground. But it's also a great enemy, and it can hurl you, tools, or equipment to the ground if you or they aren't securely attached, potentially injuring you and/or your ground crew. To avoid these situations, be sure to use the climbing and attachment gear I describe in the section "Gathering gear for a safe climb" later in this chapter.

Along with gravity, another factor you need to consider is weight. Dropping a wrench from your waist to your foot hurts, but the wrench probably doesn't weigh enough to cause serious damage. But a small home-scale wind generator may weigh 75 to 150 pounds. (See Chapter 13 for more on generator size.) Dropping that much weight from your waist to your foot may have enough force to break your ankle. Now imagine the amount of force involved if you were to drop such a generator from a 120-foot tower — you're talking about a seriously dangerous situation.

Don't take weight and gravity lightly (pun intended!). Make sure you or someone on your team has skills and experience in rigging, lifting, and handling heavy objects, or find a mentor who can help you get some hands-on training (see Chapter 12 for more on finding wind energy professionals).

Dealing with the combined weight of the system parts and the tower often means you need to work with a crane, which brings its own hazards of weight, movement, and human operator. See the section "Staying Safe when Working with a Crane" for more on special crane considerations.

Weather

Weather can make working conditions difficult for any outdoor workers, even on the ground. For tower work, weather is a major factor that can make or break the project for a day or a week.

For example, excavation for tower foundations and transmission lines requires relatively dry soil conditions; if the ground is too wet, your foundation hole may not hold its shape, and the soil will be difficult to compact properly after your concrete pour. Tower installation and climbing is ideal in dry conditions and in calm or light wind; otherwise, your whole job — communicating, securing footing, handling tools, and so on — will be more difficult. Lifting or tilting up a tower in breezes greater than 20 mph is ill advised; adding the force of wind gusts to an already challenging process may be the last straw, and you may end up with a tower failure, or at least a stressful time.

Keep an eye on the weather forecast as you're planning your tower work and factor it in accordingly. If the weather isn't safe on your work day, postpone until the conditions are more appropriate — forcing the issue just isn't worth the safety risk.

Mechanical

Mechanical hazards are those caused by spinning and yawing wind generators, and by your system of exchanging tools back and forth from ground to tower. You and your ground crew can be injured in the process of lifting heavy objects or running tools up and down the tower, and rough or inexperienced tool use can lead to damaged hands, feet, and other body parts.

For a wind-electric system, you're dealing with heavy, spinning equipment a long way above the ground. Though spinning is often one of the features that attracts people to wind-electric systems, it's also a hazard in itself. Working near an operating wind turbine puts you at risk for breaking an arm or losing your life. Any wind turbine worth buying has some means of braking it for maintenance and repair work — running a wind generator with climbers at the tower top is just too risky.

A wise wind technician doesn't try to work on rotating equipment; some who weren't so wise are no longer here to talk about their experiences.

Our small industry has a few stories — and that's a few too many — of technicians being killed by spinning wind generators, usually because they put their own safety aside in an attempt to save a generator that was spinning too quickly. This catastrophe should never happen: Your life safety should be your number one priority. Unless you have a positive remote means of braking, just walk away from an overspeeding wind generator and pick up the pieces later — it's better than not walking away at all.

Electrical

Virtually all electrical systems have shock hazards and fire hazards — these risks are inherent in moving electrical energy from sources, through wires, and to the electrical loads. Understanding how electricity works (see Chapter 4) and how proper electrical wiring can save you from the hazards can reduce your risk of electrical hazards. Electricity is an invisible hazard, so electrical meters can help you "see" it and protect you from the dangers.

Take the following electrical hazards of your wind-electric system seriously. If you get flippant or careless, you can cause a serious house fire, mild to serious shocks, or even death by electrocution.

Batteries

If you have a batteryless wind-electric system, the dangers in your electrical room are behind metal covers. If you keep your hands and screwdrivers in your pockets, you're not at risk (although you need to be aware of high and low voltage hazards during installation).

With a battery-based system, however, you have the same hazards of a batteryless system and more. Batteries are always on — unlike your electric toothbrush, they have no on-off switch. You can isolate the wiring from the batteries with disconnects (breakers) or fuses, but the battery terminals will still always be live. Treating any kind of battery with great respect will make your experience safe and productive instead of giving you (or your loved ones) stories to tell.

A critical tool when you're working with batteries, and with electrical circuits in general, is a *digital multimeter* (DMM). To look at where your wiring is seeing battery power, use the voltage function on your DMM's dial and touch your probes to positive and negative wires, wire terminations, or battery terminals. If you see voltage, approach with caution. See Chapter 4 for more on using a DMM.

Batteries present a number of hazards, including corrosive gases, acid that can eat into your clothes and skin and damage your eyes, and the potential for sparks, fires, and, with a source of ignition (such as a spark or a lit cigarette), explosion. They can also deliver very high currents, as well as create super hot arcs if you accidentally short circuit them.

When designing a battery-based system, plan for safety:

- ✔ **Design your system so that battery cables exit their box low.** This way, you avoid gas migration into the electronics, which can damage the electronics or lead to an explosion.

- ✔ **Install your battery bank away from living areas in a closed box.** This way, corrosive and explosive gases don't find their way into your home. Flip to Chapter 15 for info on battery banks.

- ✔ **Vent your sealed battery box to the outside at the highest point in the box.** Venting the box prevents explosive and corrosive gases from accumulating or migrating to electronics.

- ✔ **Store safety gear near batteries.** This equipment should include protective clothing, baking soda to neutralize acid, and an ample supply of water for flushing eyes and skin.

When working around batteries, observe these safety practices:

- **Wear eye protection, gloves, and old clothing when working on battery banks.** These simple precautions help protect you (and your nice duds) from acid and sparks.

- **Disconnect batteries before working on system.** First disable all charging sources such as wind turbines, even if there is no wind. Then isolate the battery with a breaker, not at the battery terminals, to avoid making a spark there. Reconnect batteries last to avoid short circuits and high-amperage situations as you work on your system wiring.

- **Never check amperage across battery terminals with your meter!** You'll either blow a fuse in the meter or flat-out destroy it.

- **Use a wiring diagram to correctly wire your battery bank.** Reading these instructions helps you avoid accidental short circuits, which can be catastrophic, causing fire and explosion if you have no circuit protection built in (for instance, if you accidentally drop or touch a wrench across two battery terminals).

- **Remove jewelry before working on batteries.** Dangling metal can cause dangerous short circuits, and you may be in the path.

- **Use rubber-handled or taped tools when working on battery bank.** Rubber handles keep you from making accidental connections via your tools.

- **Keep all sources of ignition, including cigarettes, lighters, sparks, and so on, away from batteries.** Batteries emit explosive gases, which you don't want to ignite!

High voltage

Anytime you're dealing with electricity, you face the risks of high voltage. Though battery voltages in typical home wind-electric systems only reach 48 volts nominal (and about 60 volts max), which may be dangerous but is not typically deadly, wind-electric systems may have a high voltage alternating current (AC) on both sides of the battery bank that can shock and kill.

More and more wind turbines are running at high voltages; 100 to 400 volts in wild (unregulated and variable voltage and frequency) AC output from the turbine isn't uncommon. The voltage may be even higher if the wind generator is running _open circuit,_ as it may be when the turbine is disconnected from the batteries and/or grid during utility outages, problems with the system, or while governing in some models. The output of your _inverter_ (DC to AC conversion equipment) may be 120 or 240 volts AC, presenting all the hazards of grid-provided household electricity. These hazards include shock, electrocution, and fire.

To guard yourself against these electrical hazards, don't touch or try to work with any wiring or connections until you've isolated them from sources of dangerous voltage. Then your DMM is your first line of defense. Using its probes and display, you can "see" what's typically invisible, verify that circuits are de-energized, and act appropriately.

All metal enclosures or other metal objects in the system should be grounded as a precaution against electric shock. Any circuits from the battery or the grid should be protected by over-current devices such as fuses or breakers to prevent fire. (See Chapters 15 and 19 for more on maintaining batteries).

Live hazards

The biggest live hazard while you're on a wind generator tower is your very own self. In the end, your ego is the biggest obstacle between you and a safe, reliable, and effective wind-electric system. Some wag said, "It isn't what you don't know that hurts you, it's what you do know that ain't so." Knowing the limitations of your knowledge is a powerful safety tool.

Live hazards also include other species, such as brothers-in-law, bees and wasps, dogs, cats, and more. I've had to deal with hornets, concern about dogs running about under our work area, and the noise from humans working in the area.

Climbing a Tower Worry-Free

If you end up with a climbable tower, you or someone you hire will need to climb it on a regular basis. Safety on the tower is an area that many people need more knowledge and training in. Most people have not had tower climbing experience prior to working on a wind-electric system, so they have little idea of just what they need to really ensure their safety while climbing. The following sections aim to give you a practical understanding of the gear and climbing technique you need to scale your tower safely and return to the ground in one piece.

 Tower work isn't something you can learn only from a book. You may be able to glean some basics from this and other books, but you also need on-the-job training. That means you need a mentor — someone to get you started in your practical understanding of the gear, climbing technique, and safety. Chapter 12 has information on finding a mentor.

Gathering gear for a safe climb

Climbing equipment is essential to climbing safety. Sure, I know some cowboys who can prove they can climb a 170-foot tower without it, and I expect doing so provides a certain thrill for adrenaline junkies. But if adrenaline is what you're after, I suggest bungee jumping or sky diving instead — at least you have some backup in those cases if you faint, slip, or just plain screw up.

Your climbing equipment is your best friend on a tower. It allows you to get comfortable, work with both hands, and use both feet to change position to maintain stability and comfort. Plus, if you do slip or lose consciousness, your safety equipment is what keeps you from plummeting back to earth. The following sections describe important climbing supplies such as harnesses, lines, attachment devices, and fall arrest mechanisms, as well as useful but optional equipment including bags and pulleys. (Figure 17-1 shows you some important pieces of climbing equipment.)

— Harness

— Fall arrest device

— Adjustable lanyard

Figure 17-1:
Harness and
climbing
gear.

— Flip line

Getting the right equipment is important. Choose carefully, considering what you need now and what you may need or want in the future — after you get used to one set of equipment, changing your habits can be difficult if and when you realize the limitations of your first choices.

Before anyone climbs, inspect all climbing, safety, rigging, and tower gear. Tighten all fasteners, check all harness and lifting gear for flaws, test any pulleys, and inspect your knots.

Having others on the crew do a second inspection on your gear is a good idea. Someone else may notice something you missed, and having each other's back is always good in this situation. You depend on each other on the tower — why not start on the ground?

Harness

A harness should become like part of your skin or clothing, fitting you snugly and giving you fall protection, a way to position and attach yourself to the tower, and a means to connect tools, supplies, and other essentials to your body while you climb and work (refer to Figure 17-1).

You can choose from many types and styles of harnesses, from rock climbing harnesses to arborist (tree-worker's) harnesses to scaffolding harnesses and more. I encourage you to look for a harness with these qualities:

- **Full body support:** This kind of harness has waist, seat, leg, and chest straps. You can't slip out of this harness if you fall unconscious, which isn't true of single-belt, belt-and-seat-strap, or even belt-and-leg-strap harnesses.

- **Seat D-rings:** *D-rings* are large metal rings (shaped more or less like the letter *D*) attached to your harness. Seat D-rings allow you to attach yourself to the tower with a sling or hardware and sit. You may be surprised at how important this ability is after working on a tower for a few hours.

- **Chest D-ring:** This D-ring attaches to your fall-arrest device; check out "100 percent fall protection" later in this section for more on fall-safety devices.

- **Hip D-rings:** No, these D-rings won't make you popular with the kids (although they *are* pretty cool). *Hip D-rings* allow you to attach adjustable lanyards and slings to your harness at your hips.

- **Other variously sized rings:** Miscellaneous small and large rings toward the back of your harness (behind your hip D-rings) let you attach lines, tools, tool bags, and so on.

Shop at safety and arborist supply houses for the harness of your dreams. Try several on, and try them with ropes, lanyards, and other gear. Sit and stand in them to simulate working on a tower — don't buy from an establishment that doesn't let you try the gear.

Lines

Lines perform various purposes in climbing: They can hold you to the tower, and they can connect your tools to your harness and you to your ground crew so that you don't have to run the risk of climbing up and down the tower multiple times for supplies. You want at least a small variety of lines for climbing and working on towers, including some or all of the following:

✔ **Climbing and rescue lines:** Arborists and rock climbers often use these lines for ascending and rappelling, and this function comes in handy as you climb your tower as well. You want at least one such line a bit more than twice that tower height so that you can rappel down the tower if necessary, or rescue an injured climber (see Figure 17-2). I prefer ½-inch-thick lines with some stretch that are soft enough to rappel with.

✔ **Small flexible lines:** These lines work as _service lines_ — a means for your ground crew to move tools and supplies from the ground to you without you having to climb up and down the tower for stuff.

✔ **Adjustable lanyards:** These lanyards (also known as flip lines; refer to Figure 17-1) allow you to attach around or through the tower and adjust your distance to find a comfortable working spot. Some (but not all) have a steel core for extra safety while working with sharp tools. I carry at least two flip lines when I'm climbing a tower and find them indispensable; I like having at least one steel core lanyard because those lines are stiffer and sometimes easier to place where you want them on the tower.

✔ **Fixed lanyards of various sorts:** _Fixed lanyards_ are less versatile, but also less expensive, than adjustable lanyards and are a useful addition to your climbing gear. I especially like a short strap lanyard with a locking hook to clip on my seat D-rings. This lanyard allows me to quickly clip into a tower strut or rung and have a comfortable means to sit and rest while climbing or working on the tower.

Figure 17-2:
A climbing crew practices taking an injured person off a tower.

✔ **Tool lanyards:** These cords can be as simple as short ¼-inch-thick lines; tie or tape them to major tools you plan to use on the tower. Spares come in handy if you're taking parts off the tower or wind generator and need a method to store or carry them.

✔ **Miscellaneous short lines:** These catchall lines let you tie off blades, assist with positioning you and your equipment on the tower, and so on.

I've been accused of looking like an octopus when I go up a tower with all these lines on my harness. I prefer that mode to getting on the tower and finding that I don't have enough lines to safely attach myself and my supplies while I climb and work. Climbing with all this stuff is an acquired skill. At first it feels a bit cumbersome, but wait out that period — you'll get used to it and end up appreciating the safety and functionality.

Attachments, pulleys, and more

A big part of tower climbing safety is making sure you're properly attached to the tower at all times. I am typically attached at least once while climbing, at least twice while stopped, and while working I often have four points of attachment. Three primary means of attaching yourself to your gear and ultimately to the tower are commonly used in tower climbing and work:

✔ **Carabiners (biners):** You use these connectors in many applications, such as attaching lanyards to your harness or hanging pulleys on the tower, and you can never seem to have enough of them. They come in various styles, with twist, screw, and button locks, or combinations of these locking methods. I *never* use non-locking biners, and I personally prefer the screw-locking variety, though I work with twist-locking biners as well.

✔ **Locking clips:** *Locking clips* are lanyard or strap ends with a hook that has a locking gate on it. You can clip them over most towers' struts or rungs to hold you in place while you rest or work. These clips come built into many adjustable flip-lines, but you can also add them to strap lanyards or other gear. Learning to open these quickly with one hand is a good thing to practice on the ground.

✔ **Knots:** Having a small repertoire of knots at your disposal proves very useful because you're likely to need knotted ropes to attach yourself and your gear to towers and tools and such as you climb. If you pick up no other knot, make sure you know the bowline, which you can use in many, many applications. It's a non-slipping, reliable knot that you can untie easily, even after it's been under a heavy strain. Next up in my list of preferred knots are the prussic, clove hitch, half hitch, and some bowline variations. www.animatedknots.com is a good online source for knot-tying instruction.

Check all your attachments before you get on the tower so you don't get any surprises mid-climb.

The gear possibilities don't stop there. The following gear is optional but really useful for making your climb easier and safer:

- ✔ **Closeable bags:** Made by arborist supply and other sources, these tough bags can hold tools, parts, and supplies, plus a bottle of water and a snack in case you need a pick-me-up on your climb. My favorite bag has a rigid bottom and rim. All should have closures, to avoid contents falling out and to the ground. (See the "Gravity" section earlier in this chapter for more on the hazards of dropped items.)

- ✔ **Pulleys:** You use pulleys for setting up lifting and lowering lines, and for moving *gin poles* (vertical temporary cranes for fixed guyed and free-standing towers; see Chapter 14) and other gear at the tower top. Split pulleys, which you put on the middle of a line so that you don't have to feed the whole line through the pulley, are best.

- ✔ **Ascending devices for ropes:** You use these devices on adjustable flip lines to adjust your position on the tower, but you can also use them for lifting and holding gear in place so you don't have to climb down for it.

100 percent fall protection

Make 100 percent fall protection your rule, and you should live to tell the tales of your tower climb. There is simply no good reason to not be attached to the tower or wind generator *all* the time!

Don't cut corners on 100 percent fall protection! Being alive at the end of the job is the number one goal. All other goals of any perceived gain in convenience, work efficiency, or anything else are secondary.

For me, having a full-time fall-arrest device when I'm climbing a tower is mandatory (refer to Figure 17-1). I just won't go up without being attached all the time, and a dedicated fall-arrest device is the simplest and safest way to accomplish that. Any climbable tower should be equipped with a fixed, vertical safety cable from the ground to as high on the tower as possible — either below the blade tips or at the base of the *yaw tube* (the highest stationary point of the tower).

Whatever type of fall-arrest device you choose, make sure it's the right size for the safety cable on your tower and that you understand how the device works. Lad-Saf brand fall-arrest devices are very common in the industry, but other cable and rope grabs are available. These devices attach to the fixed vertical cable and to the climber's harness with a carabiner. If the climber slips, faints, or becomes unable to hold himself on the tower, the device catches him by the chest D-ring, holding him safely until recovery or rescue occurs.

Test your fall-arrest device *every* time you climb the tower by stepping up the tower a few feet, running the fall-arrest device up as high as it can go, and letting yourself hang on it.

Climbing technique

You can have all the gear in the world, but if you climb a tower as if it's roller derby or a chance for bragging rights, you won't be safe. Cellphone tower climbers I know wear a sticker on their hard hats commemorating the last person in their industry who died while climbing. Unfortunately, it's not usually a distant memory, and they can often tell you the grim details.

It's almost always a lack of caution that injures or kills tower climbers. Satisfy your need for daring adrenaline-rushing activity somewhere where gravity and stupidity don't have such a high propensity for mixing.

Climbing towers is strenuous physical activity. If you're new to it, you'll be sore when you get back to the ground. Being in good general physical health helps, as does heeding some basic climbing techniques:

- ✔ **Push yourself up the tower with your legs instead of pulling up with your arms.** Climbing uses your legs and your arms. New climbers have the tendency to pull themselves up the tower with their arms, but this technique leads to sore arms before they even get to the top (where they need their arms for work). Instead, try to *push* yourself up the tower with your legs, which are larger, stronger, and used to this job.

- ✔ **Don't clutch the tower tightly and hold your body close to it.** This practice also leads to sore arms in short order. Try to find a comfortable place to be on the tower as you climb, with your arms bent somewhere between tightly clutching and fully extended. Figure 17-3 illustrates this positioning.

Figure 17-3:
Climbing intelligently leaves you stronger and safer.

✔ **Rest when needed as you climb, work on, and descend the tower.** It's not a race! On a typical 160-foot tower, I rest at least twice on the way up and at least once on the way down. I clip in, get a drink of water and maybe a snack, and enjoy the view. Climb slowly and methodically, conserving your energy for the tower work and the climb back to earth.

Letting Your Ground Crew Help You Out

The folks on the tower get all the photographs and accolades, but they can't do their job effectively without people on the ground. A ground crew's job is to make the tower climber's job as easy as possible. The *ground crew* is in charge of gathering and prepping the tools and equipment for the tower workers and making sure it gets to them in a safe and timely way. Ground crews should anticipate the tower climbers' needs, watching carefully — even with binoculars if necessary — and making sure climbers don't wait around for supplies.

The best ground crew personnel are climbers themselves — people who understand what the job is aloft because they've been there. They know that tower workers shouldn't need to lift anything significant — the tower worker has enough arm work getting himself up and down the tower. The ground crew and their service lines should put tools and equipment within arm's reach of the worker on the tower and make sure they've rigged safely for lifting big items like tower sections and the wind generator (see Figure 17-4 for rigging details). This method isn't only the most convenient for the tower climber, but also the safest strategy.

On-tower work is dangerous, but the folks on the ground have serious safety concerns as well. Falling items can injure or kill (flip to the "Gravity" section earlier in this chapter for more on these dangers).

Attempting to do tower work without a ground crew is stupid and dangerous. Stupidity is the main factor that leads to tower accidents, and it's squared or cubed if you go it alone on a tower.

Staying Safe when Working with a Crane

Crane use and safety is a big subject, and not one that can easily be covered in a book. Many jobs are hard to do without a crane. A good crane operator can "hand" you large, heavy pieces from the ground and put them within a few inches of where you need them at 100-plus feet. Working with a good crane operator is a delight. A bad crane operator, on the other hand, can make your job hard and dangerous by making you move things manually that the crane should handle, or making fast, rough, or unpredictable movements.

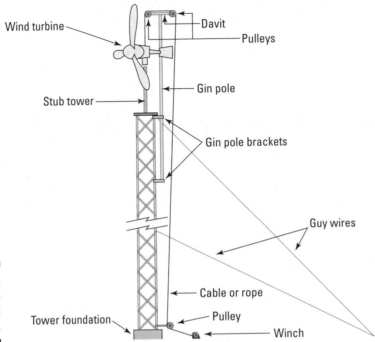

Wind turbine

Davit

Pulleys

Gin pole

Stub tower

Gin pole brackets

Guy wires

Cable or rope

Pulley

Tower foundation

Winch

Figure 17-4:
Rigging to
lift a wind
generator.

Use a respected and reputable crane company, and get references from others in the area. Talk with the salespeople, and let them know that you're new to this and would like to have their most experienced operator if possible. Befriend your operator before you get started to make sure everyone understands the job, the limitations of the equipment, and the communication and safety parameters. Knowing the hand signals that the operator uses facilitates good communication, which is half the battle. (Check out the "Using Forethought and Communication" section later in this chapter for more on hand signals and communication.) Above all, *take your time*. No crane job is worth doing quickly or in a slapdash manner. Be methodical and careful, and cranes can save you much grief and strain.

Taking No Chances with Tilt-up Towers

Tilt-up towers (which I discuss in detail in Chapter 14) bring their own set of safety issues with them. Though you aren't up in the air to work on them, they're up in the air over you, both in the lifting process and during system operation. Though these towers are very convenient, lifting them can be hazardous, especially the first time. Improperly assembled or adjusted tilt-up towers can collapse while you are lifting them, destroying your investment and posing serious dangers to you and your crew. The first time you lift your tower, be sure to read the manual for your specific tower and find someone experienced at lifting to help you get started. Here are a couple of steps you can take to keep everyone safe during your lift:

- ✔ **Appoint one leader to explain the plan to the crew and issue all instructions.** But let your whole crew know that anyone can stop the job if something seems wrong and that instructions to stop lifting must be obeyed at once.

- ✔ **While raising and lowering the tower, keep people out of the _fall zone_ where the tower may land in the event of a mishap.** The tower probably isn't going to fall, but why take chances?

After you've lifted and tuned (with properly adjusted guy wires) a tilt-up initially, the dangers decrease, and you get used to raising and lowering these engineering marvels. But you still have to watch out for dangers associated with guy wire tension, mechanical pulling devices, vehicles, and of course, human error and/or stupidity.

As with all tower work, my advice here is to take it slow! There is no reward for doing a job faster than is safe.

Using Forethought and Communication

Rules, regulations, and formal procedures exist for wind-electric systems, and if you're subject to safety enforcement agencies, you have to figure out how to comply with agency requirements. But don't delude yourself into thinking that this compliance makes you safe. Safety enforcement agency requirements may help, and compliance means you avoid fines, but no amount of bureaucracy, forms, or meetings can make you safe. Actual safety is in your hands, and primarily in your attitude, training, and habits, not in the forms you fill out, tests you can pass, or certifications you have. Whether you have to deal with the bureaucracy or not, two basic safety modes sum up real-world safety for me: think and communicate!

Thinking before you act

Wearing a hard hat doesn't mean a lot if you don't have much to protect in the first place. Your number one piece of safety gear is on your shoulders. You need brains, determination, knowledge, and experience to be safe.

Before you do any job on a tower, walk through the procedure in your mind, looking at each step for what tools and gear you'll need, what hazards you'll encounter, and what your plan is. After you've walked it through, see what your partners think. They may see flaws in your plan, or think of other gear or tools you need. Pooling your mental resources results in a safer and more efficient project.

Communicating with your crew

Communication is critical to a safe work site. Understanding what the people around you are about to do may be interesting when you're on the ground, but it's vital when you're 175 feet in the air. You have to know what to expect from your climbing partners and your ground crew.

Communication modes include

- **Voice:** Talking or shouting between climbers and between tower top and ground can be a very convenient way to communicate. However, this option breaks down when it's windy or loud or when you're on a very tall tower.

- **Radios and cellphones:** These devices can enhance safety by facilitating communication. After awhile, yelling lengthy instructions or information can take a toll on your voice and wear you out; having a convenient communication method helps you avoid these problems. Just make sure you have appropriate reception in your tower area.

- **Hand signals:** Knowledge of hand signals is an important tool to carry in your climber's brain. They come in especially handy when working with cranes. Some signals are universal — like a closed fist held out away from your body means "Stop!" — but others vary with specific crews, climbers, or equipment operators. Come to an agreement with the people you work with (before you're on the tower) so that everyone understands each other's signals.

I often use a hybrid method of communication by having a radio turned on in my pocket. My ground crew can talk to me on the radio, and I can reply with a shout or hand motion so that I don't have to get my radio out. I can use the radio when I need to have longer conversations with the ground crew.

I use what I call *baby talk* when I'm on a tower with others or even by myself. Before I do something, I say what I'm going to do: "I'm going to move my lanyard up above these rungs next; I'll need you to lift your right foot." This practice allows me and my co-workers to understand what I'm planning to do before I do it. They can be prepared for my action and also can help me improve my plan.

Chapter 18

Installation Time

In This Chapter

▶ Setting a strong foundation

▶ Raising your tower

▶ Putting in the turbine

▶ Securing your electricity-transmission system

▶ Understanding balance of systems (BOS)

*A*fter the initial planning, design, and purchase of your wind-electric system, you need to get down to the nuts and bolts of installing it. Installation is a big job — actually, it's many jobs. If you're contemplating doing it by yourself, you have to take on these roles:

✔ Excavation contractor

✔ Concrete contractor

✔ Tower assembler, installer, and climber

✔ Electrician

None of these jobs is easy or simple. All take many years of training and experience to master, and even then every new job still brings challenges. No single course or experience, let alone a beginner's book, can prepare you for your specific job. I can only hope here to raise the primary issues, terms, and procedures, and help you to know a bit about what you don't know.

Wind-electric system installation involves many skills, and it's an unusual person who has a strong handle on all these skills. More than likely, you need contractors, partners, or mentors to make a safe, functional, and reliable installation. Going it alone is a big job, with many pitfalls, so get the help you need from the start. See Chapter 12 for more on finding the right professionals for your job.

Laying the Groundwork

All wind-electric systems have tower foundations. Some may also have pow-erhouse foundations, which are a very straightforward design task — con-ventional building engineering specifies a *perimeter foundation* (common for many buildings) or slab for a small building. Engineering a foundation for a wind generator tower is a more complex job you don't find in a home con-struction manual or in general rules of thumb.

Each individual tower has specific engineering for its specific design. If your supplier can't provide clear specifications for the foundation, consider other suppliers. Tower engineers get the big bucks because they design to avoid lawsuits over damaged towers, turbines, homes, and people, so if the people you're working with can't give you the information or drawings you need, find someone who can, because foundation engineering isn't a place to ad-lib.

I speak from experience. I had an anchor for a homemade tower pull out of the ground in very wet conditions and 80+ mph winds because I hadn't engi-neered it properly. Thinking the foundation was big enough didn't save my tower and turbine from somewhat catastrophic damage — I should've been more exact.

Your finished foundation incorporates all the details covered in the follow-ing sections and provides the means to hold your wind generator up high in the sky. (Figure 18-1 shows one example, for a freestanding tower.) Make this foundation solid, and follow it with similarly solid equipment and installation, so that your system will last for decades.

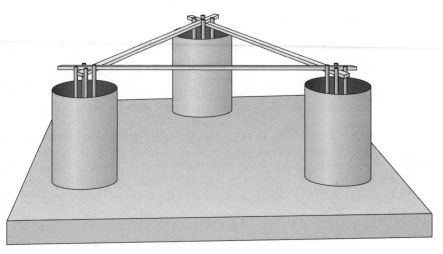

Figure 18-1:
Forming
for the
foundation
of a free-
standing
tower.

Siting and layout

Siting in this context refers to laying out the specific spots for excavation. It's different from the bigger-picture siting (picking an appropriate tower height and spot on your property) I cover in Chapter 8.

After you have the tower site chosen, you have to decide where to dig the hole or holes for the tower and anchors. Depending on your tower style, you may have to dig one, four, or five holes (see Chapter 14 for more on the different tower styles and illustrations of their footprints):

- ✔ Tilt-up towers require one tower foundation and four more for the *guy* (tower-supporting cables) anchors.
- ✔ Fixed-guyed towers usually have one tower foundation and three for the guy anchors.
- ✔ Freestanding towers need a single large excavation and concrete pour.

The three general tower styles mean there are really three different siting and layout procedures. Though the procedures for designing and siting all three tower types overlap, the following sections look at each separately.

Tilt-up

Tilt-up towers have four guy anchors, so you make five excavations and concrete pours total. The tower base pour will likely be the smallest, with a hole and foundation that may be similar in size to that of a fixed guyed tower's foundation (see the following section). The four guy anchors and the *guy radius* (distance from tower base to guy anchor) are engineered based on the *thrust loads* (horizontal wind force) on the tower. Each tower is designed to carry a certain swept area of wind turbine (see Chapters 13 and 14).

With a tilt-up tower, the positioning of the anchors is especially important, particularly that of the two side anchors relative to the tower base. Ideally, you position the two side anchors a few inches toward the lowering side of the base for the proper tension when raising and lowering the tower (the guy wires should get a bit looser as the tower is lowered and should tighten to the proper tension when the tower is raised). The four anchors should be at 90-degree intervals around the tower base. Poor positioning may lead to a tower that has guy wires becoming too tight or too loose when raising or lowering.

Be sure to use anchors that are suitable for your soil. Anchors that work well on one site may pull out on another, wetter one.

Fixed-guyed

Fixed-guyed towers typically have four excavations. The tower base foundation is often a modest hole, perhaps 4 to 6 feet deep. Its job is only to support the weight of the tower — the guy wires keep it from falling down. The guy anchor holes are usually much larger, using the depth of the concrete and the soil on top of and surrounding it as part of the holding power.

The three guy anchor holes are at 120-degree angles from each other, spaced evenly around the tower base. Your tower engineering specs tell you the guy radius; this number is the distance between the tower base excavation and the guy anchor excavations. The locations of the holes and anchors are important, but the exact height of the anchor head in the ground is not.

Freestanding

A freestanding tower has the simplest siting of all the styles. Most freestanding towers (both the tubular monopole and the freestanding lattice varieties) have one big foundation, though I've seen freestanding lattice towers (like the Eiffel Tower) with three independent excavations and concrete pours for the three legs. Sometimes this single foundation is an underground slab with a pedestal sticking up above ground level and backfilled dirt and rocks over the slab. Or it may be a single block of concrete that sticks out of the ground.

In either case, you need to follow the engineering drawings and lay out the edges of the hole. Using common building practice, you can set up stakes and corner boards to hang strings from so that your excavator operator can work to the lines. Double-check the planned hole's dimensions and the distances to the power shed and other places you have to run trenches, conduit, and wire so that you don't end up short.

Can you dig it? Excavating the area

Methods of excavation depend on the soil conditions. A backhoe or track hoe can do the job in most locations, and the smaller the machine and its bucket, the cleaner a hole you get. Backhoe operators range from artists to butchers — try to find someone with finesse, or you may end up spending a lot of money on unnecessary concrete. If you're in hard soil or rock, you may need more than a backhoe. I've seen projects where the soil geologist said that the rock was too fractured to pin a tower to, only to have the excavator take ten days to hammer a hole out of the rock with a special, machine-mounted ram.

Ideally, the hole you dig serves as your concrete form, avoiding wooden forms most or all of the time. This scenario is most attainable with freestanding towers because the hole is large enough to fit a backhoe bucket in, so you're less likely to have to oversize the hole to get the proper depth. However, depending on your situation you may need to install wooden forms with any tower type to contain the poured concrete before putting in reinforcing bar (rebar) and pouring concrete.

Following are some recommendations for making the most of the excavation phase, depending on the needs of your project and site:

✔ Plan ahead for conduit ditching and any hole(s) you may need for your power shed foundation; while you've got the machine going, you may as well take care of all your digging. Also consider what kind of access you need for concrete trucks and cranes. You may want to do a bit of road building, leveling, or filling while you have equipment on site.

✔ Ask your operator to put all the soil in as compact a pile as possible close to the hole. You need to put it all back in later, and you don't want it scattered to the winds.

Laying reinforcing rod and anchors

Regardless of your tower type, your goal in pouring your tower foundation and anchors is to reliably hold up the tower for decades. You want your concrete to hold together and be strong, which means standard reinforcing rod — commonly called rebar — crisscrossing the concrete to the engineer's specifications. Sometimes this setup can look like more steel than concrete, especially with freestanding towers, but that's just the appearance. Follow the engineering specs to the letter.

All of the rebar is tied in a grid and connected to the anchor steel. All steel needs to be suspended so that it doesn't touch the sides of your hole or forms — you don't want the steel in contact with the soil after the pour because it can rust and corrode, compromising your foundation over the decades.

The steel guy anchors connecting your concrete with your guy wires must last as well, and they're subject to more stress and weather than the anchor concrete. Typically, guy anchors consist of galvanized steel angle or rod that runs from a few feet above ground down into the center of the concrete blocks. Horizontal members are welded or bolted on, and these extend into the block of concrete and are tied to the rebar.

All anchor steel needs to be held firmly in place before you begin to pour concrete so that it doesn't move as you pour; you want your anchors to end up cast in concrete in the right places.

Pouring concrete

An uneducated spectator may think that pouring concrete is little more than pitching a bunch of mud in the ground. Though it's not rocket science, pouring concrete is a bit more involved than that. With the proper preparation (and precautions — rubber boots, rubber gloves, old clothes, and eye protection are in order), you can certainly supervise this part of the project.

Make sure all your excavation and any necessary forming are in order. You need sturdy, well-secured forming. Don't underestimate the force of a flowing mass of sand, rocks, cement, and water. It can push your flimsy wood forming out of the way and proceed to the destination of its choice.

Double-check your measurements before you order concrete to make sure they're according to the engineering plans. Your tower engineer will specify what grade of concrete you need, as well as any additives — be sure you get the right stuff. When you pour, you want to not only get the concrete in the hole but also get it to fill all the corners and completely cover all the steel with no pockets or voids. The best way to accomplish this fill is with a *vibrator,* an electrically powered mechanical device that wiggles the concrete to level it and get it into every crack and crevice. Using this device loosens the mix, which is even more of a reason to make sure your forms are tough — loose, wiggly concrete can have its way more easily with forms, where stiff, stationary concrete may not burst them.

For your tower base pour, you want to do a neat finish job with a concrete trowel for aesthetics because your concrete shows over the ground. You may want to put your initials, your sweetheart's name, or a clever or sentimental phrase in the surface to commemorate the moment. On one recent job, the owner used *Carpe Ventum!* ("seize the wind!"). A more practical thing to write would be the installation date and the phone number of the person responsible for maintenance.

Backfilling the holes

After the concrete sets but before you backfill your holes, remove any forms you've used. You want the soil to contact the concrete — that's part of the holding power of most foundation designs.

Fill the holes back up to the surface and a bit farther to allow for settling. Some tower engineers require compacting the soil, perhaps as often as every six inches, to force the settling upfront. Using a gas-powered plate or jumping-jack compactor, which you can usually rent at your local hardware or tool rental store, is most effective for this task. If you were careful with the excavation, you had the topsoil set to one side, and you can put it back where it came from so that you can grow grass again.

Towering Over the Land: Getting the Tower Up

Installing your wind generator tower is the fun part, if you ask me. It can also be one of the biggest parts of the job, depending on the height and style of your tower. It's not unusual for installation to take two or three days: a full day or two of assembly with an experienced crew, and then another day for raising and securing the tower.

Whether you're installing a tilt-up tower, a fixed guyed tower, or a freestanding tower, this stage is where the danger starts to increase. Focus on safety upfront by being aware of the hazards and having the appropriate safety gear and attitude. See Chapter 17 for more on safety.

Starting with tower assembly

Tower assembly technique depends on the tower style you're using. (Flip to Chapter 14 for more on tower styles and varieties.) All towers should have a manual; start by inventorying and sorting your parts, reading the manual, and gathering the appropriate tools and help so that you can proceed with the job undistracted by missing gear, lack of adequate help, or misunderstandings. Basic approaches for each tower style are as follows:

✔ **Tilt-up:** Tilt-up towers are usually tubular, but some are triangular lattice. The assembly procedures are similar, with variations for specific models, but follow these general steps:

1. **Lay out the tower tube or lattice with couplers and hardware (tube) or bolts (triangular lattice).**

2. **Join the sections with the appropriate hardware.**

3. **Attach the guy wires to the tower (after cutting them to length, if necessary).**

4. **Lay out the gin pole and couplers.**

5. **Attach the tower and the gin pole to hinge.**

6. **Attach the guy wires to the anchors, and the temporary guy wires or ropes to the gin pole.**

7. **Raise the gin pole, using people, props, and lifting line.**

8. **Set up your block-and-tackle lifting gear and lifting device — wind, grip hoist, come-along, or vehicle.**

✔ **Fixed guyed:** Fixed guyed towers have a similar procedure, though simpler; Figure 18-2 is an example of a guyed lattice tower assembly:

1. **Lay out the tower sections and bolts.**

2. **Join the sections with the appropriate hardware.**

3. **Attach the guy wires to the tower (after cutting them to length, if necessary).**

4. **Set up the crane or gin pole for lift (see "Lifting with a crane, gin pole, or tilt" later in this chapter).**

✔ **Freestanding:** Freestanding towers are simpler still:

1. **Lay out the tower pieces and bolts.**
2. **Join the pieces with bolts.**
3. **Set up crane or gin pole for lift.**

Tools for all tower styles are similar:

✔ Wrenches and sockets

✔ *Spuds* and *drift pins* (tapered steel tools) for fitting holes

✔ Files and wire brushes for cleaning holes and surfaces

✔ *Torque wrench* (for measuring bolt/nut tightness), if specified

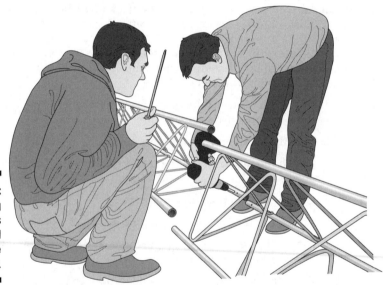

Figure 18-2: Assembling sections of a guyed lattice tower.

Hooking up guy wires

Tilt-up and fixed-guyed towers have guy wires. These wires keep the tower from falling down, so care and attention to detail are crucial for a successful installation. Some towers or tower kits come with pre-cut guy cables, which make your life easy as long as you have a standard-sized site or the manufacturer has included extra length in the cut lines. If you have to cut your own, use the specs in the manual or apply the Pythagorean theorem: Tower height squared plus the guy radius squared equals the guy length squared. Or you can lay out the triangle on the ground, using a line at right angles to the tower base to represent the ground level, cutting the guys to length in place on the diagonal line with a cable cutter, grinder, or torch.

Guy wires are either flexible aircraft-type cable or *guy strand,* such as that used by the utilities to hold up their poles. Guy strand has fewer strands and is quite stiff to work with, though effective after it's installed.

Use gloves and eye protection while working with guy wires.

You can make connections at each end of the guy wire in a couple of ways. *Swaged ends* (pre-pressed eyes) are possible on one end (usually the tower end), but aren't terribly common because they require a press and special fittings. Two connection types are most common (see Figure 18-3):

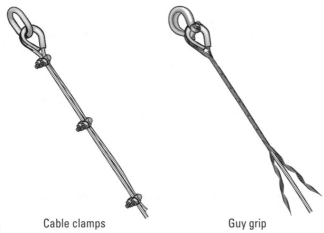

Figure 18-3: Guy termination options.

Cable clamps   Guy grip

✔ **Cable clamp:** A cable clamp has four parts:

- A U-bolt

- A *saddle* with two holes to go over the U-bolt

- Two nuts for the U-bolt

When you turn a cable back on itself at the end of the cable, you have a *live end* that goes to the working load (up to the tower or down to the anchor) and a *dead end* that is just the tail end of the cable. Cable clamps grab the live and dead ends of the cable and attach them solidly together. When you install a cable clamp, put the U-bolt side on the dead end, and the saddle on the live end — never put the saddle on the dead end.

✔ **Guy grips:** Guy grips are preformed eyes; these metal pieces spiral wrap on a cable, making an eye where needed. I find these very handy.

A *thimble* is a teardrop-shaped metal device that allows a cable to turn back on itself and make an eye without crimping the cable. Always use a thimble, whether you're working with cable clamps or guy grips; I recommend using the heaviest ones you can find, not the cheap pot-metal variety.

Roll guy wire off the reel by *unrolling* it, not by pulling the wire off the side of the spool or roll, which puts a twist in the cable.

Lifting with a crane, gin pole, or tilt

When you've got your tower assembled and any necessary guy wires attached, you can choose from three basic methods of getting your tower from horizontal to vertical:

- ✓ **Crane:** The quickest lifting method is with a crane. It can look costly upfront, but compared to paying for labor for other methods, cranes are really quite economical. Using a crane is also the safest way to lift a tower. Having a good crane operator is essential — with any machine, you run the risk of human error or mechanical failure, so start with the most knowledgeable, competent operator you can find. Chapter 17 discusses the safety concerns involved in using a crane.

- ✓ **Gin pole:** One method for erecting fixed-guyed lattice towers is clamping a vertical gin pole to the first vertical tower section and moving it up to succeeding sections as you go. The gin pole acts as a temporary crane, allowing you to lift sections. It's a slow and laborious process, but it works well if you have more time than money, or where you can't get a crane on site.

- ✓ **Tilt-up:** Tilt-ups are a different category, with their lifting method built into the tower design. After setup, they can be very convenient.

All these methods require skills and experience that you can only get in the field. An old saw says, "Good judgment comes from experience, which comes from bad judgment." I recently heard an addition: ". . . if you survive." If you're headed toward lifting your own tower, find someone to learn with for a few jobs, or take some hands-on workshops to gain experience and expertise before you take on your own project (as I recommend in Chapter 12).

Tending to Wind Generator Installation

All your work up to this point serves to get the main actor up into its limelight — the wind generator needs to be able to capture the fuel, and the fuel is high above the treetops. This stage in the construction is sometimes a bit anti-climactic because assembling and installing the turbine is usually not that big a job in itself, compared to the tower work.

The procedure is similar on the ground and in the air — it just takes you several times longer in the air. I encourage you to do as much assembly and installation as possible on the ground, though sometimes you have no choice but to do it on tower. The following sections give you some guidelines for each method of installation.

The procedures in these sections are pretty general; always follow the manufacturer's instructions for whatever specific wind generator you have.

On the ground

Assembling and attaching your wind generator to the tower on the ground is much easier, faster, and safer than trying to do it up in the air. Plus, you're more likely to do a good job on the ground — cutting corners is really tempting if you're short on tools or supplies up on a tower.

To assemble and install your wind generator on the ground (see Figure 18-4), follow this rough procedure:

1. **Gather all parts and tools at the top of the horizontal tower.**

2. **Lift and prop up the tower high enough to install the generator without its blades and tail.**

3. **Attach the wind generator, which is sometimes in multiple pieces — yaw housing, generator, and so on.**

Figure 18-4: Installing a wind generator on the ground.

4. **Lift and prop up the tower a bit higher — enough for the blades and tail to clear the ground.**

5. **Install the blades and tail, which may require a ladder to reach on larger machines or even a crane to lift parts for even larger machines (though the tower top is propped six to ten feet above the ground).**

6. **Using your manual's instructions, make electrical connections, brake cable connections, and so on.**

In the air

Although similar to the procedure outlined in the preceding section, assembling and installing the turbine in the air may take a little longer or much longer than the same job done on the ground, depending on the circumstances. If you can't install it on the ground, try to assemble the complete machine on the ground on a stand and then lift it to the tower top.

Some installers prefer to lift the turbine without the blades because of concerns about damaging them. I prefer to lift complete machines and take special care of the blades while lifting, using *taglines* (ropes attached to the wind generator and held by ground crew) to keep them away from tower and guy wires (see Figure 18-5).

Figure 18-5:
Installing
a wind
generator
in the air.

Lifting the complete machine leaves you minimal work on the tower: You can make this lift with a crane or gin pole and then just make the mechanical and electrical connections while you're in the air. Mechanically, you have to bolt the turbine to the tower, typically with flanges on each part that match up. The electrical connections may be much trickier on the tower compared to on the ground. Ideally, you have long electrical pigtails that you can push down the *stub tower* (short tubular tower section above lattice), allowing you to make your connections easily in a junction box just below the stub tower.

If you choose to lift in smaller pieces, you have more work on-tower handling individual blades, bolts, and other parts and hardware. Again, if you can arrange to do all this work on the ground, do it — you won't regret it!

Electrical Considerations from the Tower to the Ground

Making electricity is job number one for your wind-electric system; moving electricity down the tower and to your loads or the grid is job number two. This task requires several components, which I cover in the following sections. (See Chapter 15 for more on the balance of systems components.)

Size your transmission system components properly. If they're too small, you run the risk of tripped breakers, overheating wires, and possible fire. If your components are oversized, you're spending more than you need to.

Wired up: Getting the right transmission wire

A good turbine manufacturer or supplier can help you with sizing the wire you need to transmit electricity through your tower. These folks often have tables on their Web sites showing appropriate wire sizes and lengths for their turbines. Use wire that's approved for your conditions — outdoors, hanging down a tall tower, perhaps direct sunlight, and so on.

Strain relief — a means of carrying the weight of the *cable* (two or more wires in one jacket) and preventing pulling on connections — is essential and often consists of one or more *kellums grips,* devices resembling those toy Chinese finger traps but with a loop to hang from. They grab the cable as it hangs, taking the strain.

You con-du-it! Protecting your wire with conduit

Most down-tower wiring on nontubular towers is in *conduit,* plastic or metal piping that protects it from the elements, from critters, and from enthusiastic tower climbers. You typically need conduit anywhere from 1 to 2 inches thick, depending on your tower height and turbine voltage.

Secure your conduit to the tower frequently. If you get any vibration in your tower — and you're likely to from time to time — you don't want to hear your conduit slapping against the tower legs. The easiest method is to install conduit on the ground, where you can slip it on the inside of the tower sections, glue it, and zip tie or clamp it to the tower. If you need to lift the tower in more than one section, use *slip-joint conduit sections,* which allow you to make a secure and weatherproof connection after you've joined the tower sections. These expansion joints accommodate expansion (lengthening) and contraction (shortening) with temperature.

Brake it up: Stopping the system when necessary

Don't break the wiring between the wind generator and the *load*, be it a battery bank or a grid-tied inverter. This situation allows your turbine to free-wheel, which may lead it to run faster than you want.

However, a *dynamic* or electrical-shorting brake is a very handy thing to have because it allows you to stop the turbine at will and works on most turbine models (but verify with your manufacturer). This device can be as simple as a common *single-throw, triple-pole switch* (one switch that controls three phases of the circuit) that is paralleled into the transmission wires and wired so that throwing the switch shorts the three phases. These switches are available at larger hardware stores.

Depending on where the *rectifiers* (which convert wild AC to DC) are in the system, you may need a second switch to disconnect the wind generator from the battery bank to avoid shorting the battery. Having one brake switch at the base of the tower and another in the power room is common.

You're grounded! Grounding the tower

Lightning is a serious concern with all towers, and it can be extremely serious if you live in a high-lightning area. At a minimum, grounding all tower parts is essential to dissipate any difference in electrical charge between tower and ground and to divert lightning away from your equipment into the ground.

Dedicated *ground wires* connect the tower base to the *ground rod* (an 8-foot copper or copper-clad rod driven into the ground). Separate wires connect each guy wire to the guy anchor ground rods. So with a fixed-guyed tower, for instance, you need at least four ground rods — one at the tower base and one at each guy anchor. In high-lightning areas, you may also connect all these wires together underground, or even have a large underground mat of copper to dissipate the charge.

Balance of Systems

Balance of systems (BOS) is a catchall term that includes all the other equipment beyond the stars like wind generator and tower — charge controller, inverter, disconnects, and so on.

This stage is where wind-electric systems can get complicated and technical, so you may need more help here depending on your skills and experience. Each different wind generator and each different system configuration has a different set of equipment with different specifications. Figure 18-6 shows a typical battery-based layout.

Lean on the manuals, your supplier, your partners in the project, and the manufacturer. Even very experienced renewable energy system installers have to familiarize themselves with each new piece of equipment they encounter. And that's not a one-time job — manufacturers regularly upgrade equipment, adding new features, fixing bugs, and changing configurations.

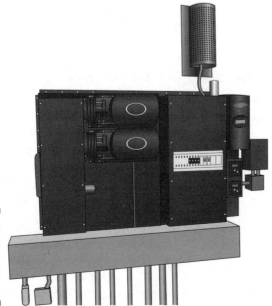

Figure 18-6: A typical battery-based BOS layout.

Inverter

An *inverter* converts direct current (DC) electricity (usually low voltage) to alternating current (AC) electricity (usually high voltage). As I mention in Chapter 15, wind-electric systems use two distinct types of inverters: battery-based and batteryless.

- ✔ *Battery-based inverters* have many more wiring and configuration possibilities, depending on the overall system design and the specific equipment you buy. If you buy a prewired system, installation *may* be as easy as screwing the boxes to the wall and connecting the DC and AC leads. More likely, you need some thought and care to tie your inverter into the system properly. See Chapter 15 for more information.
- ✔ With batteryless systems, installation is more likely to be simple, with DC leads leaving the rectifiers (and possible voltage clamp) and coming into the inverter, and grid-bound AC leads heading out of the inverter.

Batteries

Batteries are one of the most dangerous portions of your electrical system. When installing batteries, your primary objective should be safety. These always-on devices have weight, shock, and chemical hazards, among others. See Chapters 15 and 17 for more on these dangers.

Battery design, specification, and installation could be the subject of a whole 'nother book. Installing them calls for slow, methodical work and recognizing the high potential for sparks and explosions if you make careless mistakes. Chapter 15 has more details, but key installation tips include

- ✔ Make a diagram first.
- ✔ Use protective clothing and rubber-handled tools.
- ✔ Lift with your legs, and a partner.
- ✔ Protect all terminal and cable ends.
- ✔ Think!

Connecting your batteries to the system *last* is the safest way to go. (But make sure your wind generator is mechanically or electrically braked until it is connected to the battery.) Your system has a main battery breaker, which will be the final connection. Do your homework so that you don't end up with sparks or worse when you flip this breaker. *Triple*-check your wiring diagram before making connections.

Charge controller

Battery-based systems have a charge controller to protect the battery from overcharge. (These "black boxes" also have many other possible functions — check out Chapter 15 for more info.) Charge controller installation can be as simple as mounting the controller box on the wall and hooking four to six wires up to it. Sometimes you find controllers integrated into breaker boxes or even inverter cabinets, with much prewiring done. Other times you have to turn to the manual to help you connect a variety of boxes and understand how they interact so you can make them "play" well together. Frequently, the installation involves programming battery charge parameters and other features, which I cover a bit more in Chapter 15.

Metering

Metering installation details depend on your local requirements and your own preferences. More and more metering is being included within inverters and charge controllers and in integrated system packages. This kind of situation requires no or minimal wiring depending on the package. You may need to wire a *shunt* (a measured-resistance device) into the negative side of the system for wind generator production metering or battery state-of-charge metering, but this element typically comes with the meter.

What the utility or incentive program requires is another piece. It's often a utility-style dial or digital meter to measure how many kilowatt-hours (kWh) of electricity you produce. This component may be installed by you, your installer, or possibly the utility. The utility will definitely install its revenue meter — the one that counts the kWh bought and sold.

All of this metering installation obviously requires electrical knowledge, skills, and sometimes credentials. Find out what it takes, and don't treat the installation lightly because you're still dealing with serious dangers.

Breakers and so on

Specifying and installing electrical equipment is work for electricians or those trained or practiced in electrical work — I don't recommend dabbling in it. I've learned from my own experiences and now have an experienced renewable energy electrician work on my own system. Electricians who don't have experience with renewable energy systems often can't provide adequate help because they have little experience with low voltage DC and may not readily grasp the whole scheme of a new and different technology.

Wind-electric systems require a variety of disconnects, *overcurrent devices* (fuses and breakers), grounding devices, and so on to ensure functionality, efficiency, and safety. Going into these nitty-gritty details is beyond the scope of this book, but that doesn't mean they're not important. In fact, they're vital parts of a complete system.

In some cases, much of this installation work may be done for you by the manufacturer if you purchase a packaged system. More of these decisions will fall onto your plate, or that of your designer/installer the farther you get from a basic, batteryless system.

Down to the wire: Understanding electrical considerations

The transmission wiring isn't the only wire in your system. Small wind-electric system wiring is different from ordinary home wiring, and even electricians get puzzled sometimes. The biggest difference is that there are multiple sources of electrical energy in the system, working at multiple voltages and frequencies or even with DC, which has no frequency.

Every circuit in the system needs to be designed with a view to the maximum voltage and the maximum amperage (current) that may arise in that circuit. In some cases, these are easy numbers to find — for example, a 48-volt battery system probably won't greatly exceed 60 volts DC. But the maximum amperage that a wind turbine will generate can be a lot harder to establish, so you may need to leave a bit more headroom in some cases. The rating of a piece of wire, a diode, a fuse, or even a connection in the circuit has to meet or exceed that rated voltage or current for that circuit. A higher rating is always good, and in many cases it's prudent to add a factor of safety to the rating. Such factors are enshrined in the code for many situations. And in most cases, a higher rating will add to the efficiency and reliability of the system.

Wiring also affects efficiency. For example, the wires need to be thick enough to prevent wasting energy, even if that waste is only a gentle warmth in a very long wire. It's important to look at the voltage drop in long wire runs, especially at low voltages where this voltage drop becomes a significant percentage of the circuit voltage. If you lose 10 percent of the voltage, then you lose 10 percent of the energy that the wires transmit. Where the amperage is likely to be very high (such as between a 12-volt battery and an inverter), you should keep the wires as short as possible and choose the thickest you can afford.

If you do any of your own wiring, take care to read the equipment manuals carefully. They've been written specifically to help with the idiosyncratic problems of small wind-electric systems. They often give detailed guidance on wire sizing and on safe installation.

And by all means, ask electricians to help you with the wiring of your renewable energy system. They'll make a tidy job. But don't necessarily expect them to have all the answers, even when they sound confident. Ask for a design document that shows how the various wire sizes and fuses were chosen. Ask for clear instructions on how to shut the system down and isolate parts of repair or maintenance. A carefully planned system is likely to be reliable and safe in operation.

Chapter 19

Living with Wind Energy

Depending on the system configuration (batteryless grid-tie, grid-tie with battery backup, or off-grid), wind-electric systems can be turnkey and automatic or very hands on. You may hire an installer to put in the system, and the company may do all the design work and all the dirty work, leaving you with a check to write and a spinning turbine to watch. Or you may be very involved in the installation and operation of the system and intimately familiar with the details.

Regardless of whether the system is turnkey or hands on, all wind-electric systems need monitoring and maintenance, both of which I discuss in this chapter. The idea that you can buy a wind generator that won't need maintenance is a total myth. And even if this machine existed, it would be very, very expensive.

Wind-electric system reliability is directly proportional to the quality of the components and design *and* to the level of monitoring and maintenance you give the system.

What Are You Looking At? Monitoring Your Wind System

Wind-electric systems are not all benefits — enjoying the weather, watching the dynamic motion of your new machine, and bragging to your neighbors. Your system is also a big responsibility. One advisor once said, "If you treat it as if it's unbreakable, it won't be." But if you treat it as if it needs regular care and maintenance, you may enjoy a long and productive relationship with your wind generator.

If your system is well designed, the daily operation of it should be fairly transparent. Your job shouldn't be like that of a refinery operator, who has to constantly monitor conditions and adjust equipment and processes accordingly. In most cases, you should be able to let it do its job while you do yours.

The primary characteristic of a good wind-electric system owner is *awareness*. Being aware of your system allows you to spot minor problems before they become major problems and to make any necessary repairs (either by yourself or with the help of a professional). And awareness helps you determine whether you're getting what you paid for: a system that reliably produces the amount of energy you'd hoped for.

The specifics of your awareness run the gamut from listening to the machine to watching a variety of meters. In this section, I describe a few particular things to keep an eye on and explain the importance of tracking your system's historical data.

Watching the wind

Having a wind generator brings an easy awareness of wind and weather. You don't have to turn on the Weather Channel to know the wind's direction, general speed, and gustiness. You can just look up (way up, I hope) at your wind generator and get the current report.

Wind awareness, however, goes beyond your sense of its being calm or windy. You can measure wind speed, both instantaneously and over time (see Chapter 8 for information on a variety of methods).

Seeing the wind through new eyes

Quantifying the emotional value of living with wind energy is very hard. Having a wind generator changes the whole way you look at and feel about the weather. BWE (before wind energy), you may be concerned about buttoning up your overcoat or retrieving the trashcans blowing down the street. After installing a wind-electric system, every breeze brings you visions of kilowatt-hours either running into your battery bank or spinning your meter backwards.

I've seen outlooks change many, many times when people first see a home-scale wind turbine: that look of admiration — the dreamy eyes and longing for a wind generator of their own. There really is nothing quite like the satisfaction of making your own electricity with the wind. It brings out the fanatic in some people. In my case, one of my wind turbines is visible from my pillow, through a skylight in our bedroom. The other two are out of direct view, but mirrors in the skylight well allow me to view them. I wonder if there's a recovery program for people like me!

Having an _anemometer_ — a wind gauge — in your system is not optional to me. How do you know what sort of performance you're getting if you don't know anything about the resource? Ideally, you'll have a recording anemometer that gives data to your computer either directly or through download. I love to have weekly, monthly, and annual averages and peaks. Here's what these wind readings tell you:

- **Average wind speed:** The average data correlated with energy production data (which I cover later in this chapter) lets you know whether you're within manufacturer or installer projections.

- **Peak wind speed:** The peak data gives you some sense of what your turbine has to deal with, and your eyes tell you how your turbine stands up to that.

Picking up on power

As you watch your turbine on a moment-to-moment basis, seeing the instantaneous energy production is fun and interesting. That's wattage, and I'd love to have an analog wattmeter, because the device's dial makes it very easy to see how the production varies. More often, a wind-energy user has a digital wattmeter, or you may have analog amp- and voltmeters with which you can do the math (because volts times amps equals watts). I describe meters in more detail in Chapters 4 and 6.

A voltmeter that goes before the _rectifiers_ (wild AC to DC conversion devices) is also a very useful device because it can show you while standing in the power room that the turbine is spinning. You watch the voltage rise up to about your battery voltage or at the cut-in of the inverter, after which the machine will start working for you. This helps you become aware of when the machine is actually generating and when it's not spinning fast enough to do its job.

You or your installer establish baseline values for your system early in its operation. For instance, you may know that the set point at which your inverter starts to "sell" electricity to the utility is at 56.2 volts, and under normal operation, the battery voltage will generally be at that level or below. With this information, you can recognize when something is amiss; if you see the battery voltage at 59.6, you know that either the grid is down or something is wrong with your inverter or settings. Similarly, being able to monitor your amperage and wind speed helps you notice small issues before they become big problems.

Examining energy

Energy, which is measured in kilowatt-hours (kWh), is the most important thing to keep track of. If you didn't know what a kilowatt-hour was before you bought a wind system, you'll find out during the design process or when the system starts working; only the most careless wind system owners don't know that the main goal of having a system is to make kilowatt-hours.

A wind generator without a kilowatt-hour meter is like a business without accounting — you don't know what you've made or lost. An encouraging thing happens when you start making your own energy: You start noticing how much energy you make and how much you use. An even better thing happens next for many people — they start to use energy more carefully so they can make the most of their precious wind-generated kilowatt-hours.

Because incentives are often based on the number of kilowatt-hours your system produces, many if not most grid-tied systems today have a dedicated production meter. If that's not the case, I encourage you to invest in a dedicated kilowatt-hour meter. You can buy reconditioned meters for less than $50, and they hardly add to your system cost if you install them when you install the system. To keep track of your on-grid energy consumption, you can add a package of sensors and a remote readout to your distribution panel to monitor your usage. If your utility company offers net metering (the ability to "sell" your surplus energy and create a credit), you can often simply check your utility bill for these numbers.

If you're off-grid or in a battery backup situation, you need to be aware of the kilowatt-hours you're using and the kilowatt-hours left in your battery bank. State-of-charge (amp-hour or watt-hour) meters do this job for you; some are included in charge controllers or power panels, and others are separate electronic components. Figure 19-1 (later in this chapter) shows a variety of these components.

Knowing how many kilowatt-hours your system generates and how many kilowatt-hours your home uses can give you a whole new perspective on the natural resources on your site and on your lifestyle. Over and over, I've seen people get more interested in using energy wisely when they can clearly see these two things. If your goal is to make all of your electricity, monitoring helps you keep track of how much energy you're using versus how much you're generating. Even if you're providing only some of your electricity, monitoring how many kWh your generator makes is key.

Listening up: Mechanical and electrical sounds

Your ears and eyes are great primary diagnostic and operational tools for your wind-electric system. If you hear something out of the ordinary, you need to try to figure out what caused the change in sound and address the cause. Small noises do come and go, but sometimes they go with a bang or a crash — and with your investment on the ground. Here are a few examples of noises associated with a system's mechanical and electrical systems:

✔ An unbalanced blade rotor may cause vibration in the wind generator and all down the tower. The unbalance may be because of poorly
matched blades, damage to one or more blades, or a problem with the
furling system adjustment.

✔ A minor increase in blade noise may indicate dirty blades covered with
bugs and crud. Time to clean!

✔ A dropped or poorly connected phase in your turbine's alternator leads
to choppy operation, if the machine runs at all. Listen to the alternator
whine early in the machine's life so you know what the alternator should
sound like.

Clunking, grinding, and rubbing noises are all very bad news. Find out in a
hurry what's causing them before you get the experience of finding your
machine digging itself an early grave.

In addition to these somewhat subjective sensory tools, your system should
have a variety of meters that can be eyes and ears to the parts of your
system that you can't see and hear. I describe a variety of these meters earlier in this chapter.

Tracking your system's historical data

Having an instantaneous view of wind and production is mostly a way to
satisfy your instantaneous curiosity. Having a historical view is a much more
useful tool, both for you and for future wind energy users in your neighborhood. With it, you can track usage and compare it with the wind resource and
with projections from manufacturers. You can compare production for various months and years, noting whether your system is still performing as it
did in the beginning.

Today, people are using sophisticated systems that log data directly to
the Internet for this historical view. Many of these systems are not really
plug-and-play, and they require a fairly sophisticated owner or installer to
get them set up and running. But after they're set up, being able to look at
present and historical data on your system from anywhere you can get an
Internet connection is fantastic! Some of these systems are built into balance
of systems (BOS) packages. Others are stand-alone. You can figure out which
is right for you by quizzing your supplier or installer.

If you're not in the market for these systems, which can cost anywhere from
$500 to $2,500 or more, you have other options. The meters in your system
components may look something like Figure 19-1. Some of these systems
have some internal historical logging, but the data rolls over and you lose it
every few months or after a year. As long as you record the data (kWh,
average wind speed, and perhaps peak wind speed, high and low voltages,
and max amperage) with a pencil, paper, and clipboard or into a spreadsheet,
you can have the information you need without the high price tag.

One of my wind charge controllers, for instance, shows me production for the last 128 days. If I'm on top of it, I can record the information and reset the meter to zero every few months. My wind data logger holds monthly average and gust data for a year before it starts to write over the data. The best strategy is to record data every January 1 and then get a fresh start.

Having a record of early data gives you a baseline of normal performance — assuming you have a good start with your system. Over time, you get a feel for what the numbers should be under different circumstances. Best case scenario: You'll know that your machine is still kicking out the kWh that you expect in a certain average wind speed. Worst case: When you see the *ammeter* (which measures an electrical current's strength) go to zero on a dark and stormy night, you may fear to go out in the morning and see what's left of your wind generator — it's happened to me, and it's no fun.

Data logger for anemometer

Ammeter

Volt meter

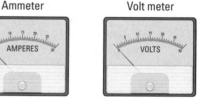

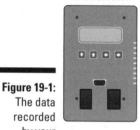

Charge controller

Battery SOC meters

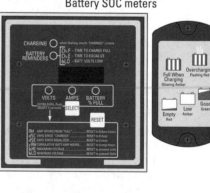

Figure 19-1: The data recorded by your system's different meters can help you understand whether it's performing as designed and help you manage the energy.

A Little TLC: Maintaining Your Wind System

Wind-electric system maintenance is crucial. It should be done at least once a year and more often on severe sites or with poorly made equipment.

Fortunately, on quality machines, maintenance is not difficult or terribly expensive in terms of supplies or replacement parts. You mostly need to be willing to spend the time or dollars to *do* the inspection and follow up on any problems. Labor is usually the largest cost: Getting an experienced crew out to lower or climb your tower and do the work will likely cost several hundred dollars. (Flip to Chapter 12 for tips on finding wind-electric professionals.)

If you did part or all of the installation, you're the very best person to maintain it. Consult with your contractor and supplier and use or make a maintenance checklist. Learn how to do it well — and safely! For safety advice — whether you're climbing a tower, topping off the fluid in your batteries, or working on the electrical equipment — see Chapter 17.

In the following sections, I describe the biggest maintenance tasks that should be performed on your system every year (or every few years). Maintenance does involve some careful and detailed work, but after you know what you need, you find that it isn't that demanding a process. What is demanding is that you get the job done on a regular basis. Put it on your favorite calendar, and if it's an electronic calendar, click on "repeat every year."

Inspecting the tower

Inspect all tower hardware before and as you climb or lower the tower for maintenance. This goes for tilt-ups, guyed lattice, and freestanding towers (all of which I discuss in Chapter 14), and it's especially important with guyed towers. Someone needs to oil moving devices (such as turnbuckles) and check the following hardware for tightness:

- ✔ Guy anchors where they leave the ground and just below ground level
- ✔ All guy anchor attachment hardware, turnbuckles, guy grips, and cable clamps
- ✔ Tower base attachment or pin
- ✔ All grounding equipment and connections
- ✔ Bolts on the tower — check them as you climb
- ✔ Guy-wire-attachment hardware on the tower
- ✔ Safety cable attachments and other climber safety and convenience hardware

Also check the tower itself and hardware for rust and corrosion. You can use spray galvanizing or rust-preventive paint for touching up bare spots on the tower to avoid rust.

Checking the wind generator

The main focus of your maintenance check is the wind generator itself. This is the spinning piece of equipment, so it shows the most wear and tear, and it has the most vibration to work the hardware loose. One bolt working itself loose can soon turn into more bolts loose and then a complete failure, so be vigilant about regular, thorough maintenance.

Maintenance tasks include the following:

✔ Tightening all hardware to factory specifications

✔ Greasing any grease *zerks* (fittings through which you can grease mechanical joints) or areas specified in the manual

✔ Inspecting blades for wear, cracks, and crud

✔ Checking blade balance and tracking

✔ Inspecting all wiring connections and strain relief

✔ Inspecting mechanical parts of the turbine, such as the furling system, yaw bearings, and main bearings (see Chapter 3 for an introduction to a generator's parts)

Assessing electrical components

In addition to the turbine's electrical components, you want to periodically check on the transmission cables down the tower, especially where they're supported and where any connections are made. At the tower base, you likely have a junction box (electrical connection box) and perhaps a brake (to electrically stop the rotation of the turbine). These are areas with connections, and connections normally loosen over time, so checking them after the first year and every few years thereafter is wise.

In your power room are many more connections and therefore many more opportunities for loose connections. Vibration is less of a factor here, but you should still inspect connections for tightness every few years.

Inverters and charge controllers are essential electronic devices, and you should satisfy yourself that they're operating properly and double check their settings during your maintenance session. At installation, recording key settings in a logbook gives you a baseline to check against periodically. Reviewing production records can also ferret out issues with these components.

Most typical electrical components — breakers, wire, grounding, distribution panels, meters, and so on — are built for the long term. Occasional failure is possible, but the most likely point of maintenance issues is the wiring connections to these devices. Physical tightening of connections every few years is in order, along with visual inspection to discover any sign of wear or heat.

Maintaining batteries

Batteries bring a whole other level of maintenance; if you choose to have batteries, don't avoid this work. Flooded batteries require the most maintenance; sealed batteries are less involved, but be sure to keep the batteries clean and the connections tight.

The following list gives you the high points of their maintenance (which you should always do while wearing protective gear, including eye protection, gloves, and corrosion-resistant clothing.) Head to Chapter 17 for more on battery safety.

- ✔ Check the battery fluid level and top them off with distilled water as necessary.
- ✔ Check battery connections for tightness and corrosion.
- ✔ Check specific gravity (battery fluid density) with a hydrometer.
- ✔ Clean battery tops.

Less prominent in a maintenance checklist is to discharge and charge batteries with care and attention. Most batteries don't want to see regular discharges below 50 percent *state of charge* (SOC). Some can take discharges to 20 percent SOC (which is the same as 80 percent DOD — *depth of discharge*), but all want quick and full recharging. One of the worst things you can do to a battery is to discharge it deeply and leave it that way for a long time.

Being aware of how much energy you're using and what your battery SOC is are regular parts of maintenance. Recharging your batteries quickly after an outage or deep discharge is important. And it's important to fully charge them on a regular basis — I prefer at least once a week, but every month or so isn't unreasonable — if you want your batteries to live a full and long life on God's green earth.

Troubleshooting and replacing batteries

Problems in battery-based wind-electric systems often end up pointing you back to the batteries. Two strategies can help you spot and identify these problems:

✔ **Stay generally aware of your battery voltage.** Voltage varies from a low of just above the nominal voltage (48 volts in most whole-house systems) to a high of about 25 percent above nominal (60 volts for 48-volt systems) in normal operation. Over a number of days, you may see voltage at 48.6 when batteries are deeply discharged or being used heavily, and you may see voltage up in the mid-50s when the batteries are full and/or under heavy charge. If you see wild swings in voltage during charge or discharge, it's a sign that your battery bank is suffering.

✔ **Periodically measure individual battery or cell voltages, looking for dead cells or an imbalance between cells.** Dead cells require replacement, and you can solve an imbalance with an *equalizing charge,* a controlled overcharge of the battery bank.

A battery failure can be seen as a cause of your problems, but to an extent, it may be only a symptom. Premature battery failure may indicate that you aren't producing as much as you're using, and so the battery is chronically in a low state of charge. It isn't easy to determine the real cause of battery troubles, but you should consider both the battery itself and the way it has been managed.

Bad connections may result in poor performance, but your routine maintenance should find and correct this. In the end, you need to replace your battery bank every five to fifteen years, depending on the quality of the bank in the first place and on how you care for it. If you buy and care well, you may get ten or more years out of a bank of batteries.

Battery watering systems: A handy time-saver

One major task in battery maintenance is adding distilled water to the battery cells. This task can be tedious and time consuming, and it may be one that you don't readily get around to. You have to remove each cap, peer down into the cell with a flashlight (no matches!), fill it to the proper level with a funnel and a jug of distilled water, and replace the caps. A typical home's battery bank may have 30 to 50 caps, so this process can get to be a pain.

Enter a *battery watering system:* Special caps replace the original battery caps. They have a float down in the electrolyte and a valve that turns on and off the flow of distilled water. The water is provided under pressure from a pump sprayer or tank about the batteries. These systems can be quite costly upfront — as much as $15 per cap. But they are *very* convenient and flat-out worth it. When it's time to water the battery bank, you simply pump up the sprayer to pressurize it or turn on the valve from the elevated reservoir. Distilled water flows to each cell, and the valve shuts off when each is full.

Part V
The Part of Tens

The 5th Wave · By Rich Tennant

"Here's an energy-efficient number I think you'll like: double-insulated turrets, radiant-heated dungeon, polyurethane foam-core drawbridge, low-flow moat..."

In this part . . .

The Part of Tens lets me give you some basics in a bite-sized form. Chapter 20 talks about ten goals for your wind-electric system, and Chapter 21 details ten mistakes you want to avoid. To top it off, Chapter 22 tells ten stories of wind-energy users to give you examples of successes and failures.

Chapter 20

Ten Essential Steps toward a Successful Wind-Electric System

In This Chapter

▶ Knowing your energy use and becoming more efficient

▶ Finding the right wind resource and spot on your land

▶ Getting a reality check before you commit

▶ Sizing your tower and generator correctly

▶ Installing and maintaining a quality system

With everything people do, they have a goal or goals. If you're going out on Friday night, for example, your goal may be to have a few laughs and enjoy some friends and music you love. Or it may be to make a business contact that will lead you to your first million dollars. You act very differently depending on your goals, or you may be disappointed in the results.

When people decide to have a wind-electric system, they have goals as well. See Chapter 2 for the list of basic goals or motivations I've seen in my clients and students; yours may fit in one or more of those categories or be different altogether. No matter what your goal is, you want to focus on some essential steps to reach it. The ten steps in this chapter top my list.

Know Your Load

Very few people I talk to want a wind generator but don't care how it performs. You measure performance for a wind generator in kilowatt-hours, and for almost everyone, the goal is to make some so you can use some. If you don't know how many kilowatt-hours you use (in other words, your *load*), you're a prime candidate for disappointment.

The average American home (without electric heat) uses about 30 kWh per day. Are you average? And do you want your expectations based on a national average? No. Find out what your daily energy usage is by doing a load analysis or at least looking at the past year of utility bills. Chapter 6 helps you look at the big picture and then go beyond and look at details.

In the end, I hope you or your system designer will be able to say something like, "This system will produce 14 kilowatt-hours per day on average, which will be about 70 percent of your home's energy use." You can't get to a statement like that without knowing your home's energy use!

Shrink Your Load

As soon as you know your daily energy load in kilowatt-hours, your very next focus should be on doing everything you can to reduce that number. This is without question the most important phase of your wind energy adventure if your goals include saving money or the planet.

Energy efficiency, or *negawatts,* as energy guru Amory Lovins calls it, is the cheapest energy you'll ever buy — the energy you never have to make. By shrinking your load, you shrink almost everything about your wind-electric system, including the following:

- Wind generator
- Tower (thrust load, and therefore strength and cost, but not height)
- Transmission wire size
- Charge controller size
- Battery bank
- Inverter
- Other electrical components

Guess what. Every one of these things costs money! So shrinking your energy load shrinks the cost of your system all the way down the line. It's not exactly linear, because some components don't shrink proportionally with the load and costs don't either. But cutting your load in half certainly cuts your system cost by a third or more. Chapter 7 has the details on increasing your home's energy efficiency.

Know Your Wind Resource

After you've determined your load and then reduced it, nothing is more important than knowing your energy resource. This is true of all resources — sunshine, wind, water, or biomass. In your case, I'm talking about how much wind energy you have on your site. For home-scale sites, you measure this figure in a simple average wind speed, and the range you see on home-scale sites is usually from the low single digits to a maximum of 14 mph. Sites worth tapping start at about 8 mph annual average, and having a home-scale wind site above 14 mph is rare. In fact, above 12 is not that common in residential environments on tall towers.

Determining your wind resource is probably the trickiest work you do in designing a wind-electric system. At the home-scale, people don't usually have enough money to do a full-blown wind study, so they rely on a number of methods, some very subjective and others only partially subjective. (See Chapter 8 for discussion of these methods.) I recommend you use as many as possible and be conservative in your estimate of your site's resource. Overestimating leads to disappointment. Underestimating leads to a pleasant surprise.

Know Your Site

Your specific site characteristics affect your wind-electric system design. The more familiar you are with your site, the better your design will be. If you have a couple of acres of flat farmland in the middle of hundreds of acres of the same, deciding what your options are for siting your tower and electrical equipment may be very easy. If you own 100 acres of varied topography, you have a more complex task.

To go into your design phase well prepared, you should know, at a minimum, the following things about your site or sites:

- ✔ Highest points
- ✔ Distances between potential sites and your home and/or the utility grid
- ✔ Tallest obstructions on the property and nearby neighbors
- ✔ Overall topography of region and how your property fits into it
- ✔ Existing or potential home sites and other energy use locations

I recommend using a detailed topographic map and drawing in the preceding features, compiling as much information on your property as possible in one visual resource. Check out Chapter 8 for additional pointers.

Be Realistic

In the past year, I've seen these wind-electric installations:

- A $15,000 installation of a 12-foot turbine on a much too short (33-foot) tower generating 250 kWh per year for a simple payback of about 220 years. Cost of energy: about $1.30 per kilowatt-hour, or about 6 times the local utility cost.

- A $30,000 installation of two 12-foot turbines on 33-foot towers below the tree line, which will likely produce only a very few kilowatt-hours and have a payback of several hundred years, if the turbines last that long. Cost of energy: $5 or more per kilowatt-hour, or at least 25 times the local utility cost.

- A $90,000 installation of a 21-foot turbine on a 60-foot tower generating only 750 kWh per year for a simple payback of 1,000+ years. Cost of energy: about $4 per kWh, or 40 times local utility cost.

Now I'm the first to say that money isn't everything. I don't even think it's the most important thing. But I don't think the buyers of these systems would have made the purchases if they'd been realistic upfront. And that means they would avoid the serious disappointments that resulted.

Do at least some rudimentary calculations of return on investment (discussed in Chapter 10) so that you know what you're getting into before you buy!

Use a Tall Tower

Being realistic requires tall towers on almost all sites. The bad examples and experiences in the preceding section were all a result of towers that were short — much too short. You can ruin your prospects for making a lot of wind energy in other ways, but putting your wind generator on too short a tower is the most common. It's a mistake that the wind-electric industry has yet to learn from entirely. It continues to have companies coming in and advertising products and installation methods that try to defy the physics of wind.

But defying physics works better in the movies than in the real world. The reality outside the TV box is that

- Wind energy increases with the cube of the wind speed (V^3)
- Wind speeds increase as you move away from the Earth and its obstructions

Tall towers, tall towers, tall towers! There's no more-important design advice that I have for you. Bend the 30/500 rule (site wind generators at least 30 feet above anything within 500 feet, minimum; see Chapters 5 and 14), and the performance of your wind-electric system will suffer. Ignore this advice, and you may end up with one of those systems with a 100+ year payback and energy that costs much more than you're paying now. Does that serve your goals?

Use a Large Rotor

 After the wind resource, the next most important factor in how much wind energy you'll get is how big your wind collector is. With wind generators, the collector is the *swept area* of the rotor, the circle that the blades describe. The bigger this collector area, the more energy you'll collect. Twice the wind generator rotor swept area gives you twice the energy capture.

I often hear people who aren't really very familiar with wind generators get fixated on designing machines that are "more efficient." Although increasing efficiency is a worthy goal, you need to remember that you're dealing with a free and abundant resource. If you want to increase the output of a machine, you can choose from these two options:

✔ Spend hundreds of thousands of dollars on research and development to increase the efficiency of your machine by 10 percent.

✔ Add 10 percent to the swept area of the machine, at the cost of perhaps tens of thousands of dollars or less in engineering changes to the design. For a 12-foot diameter machine, increasing the collector area by 10 percent amounts to adding 4 inches to each blade. Doubling the diameter gives you four times the swept area.

Flip to Chapters 5 and 13 for more information on wind generators and the importance of a large swept area.

Buy Quality Equipment

Dead and broken wind generators generate no electricity. Reliability must be design parameters numbers one, two, and three. If you put efficiency, cost, speed, aesthetics, or any other parameter ahead of reliability, you probably won't achieve your goals, whatever they are. I've never met anyone whose goal is to spend thousands of their dollars on a broken piece of equipment on a 120-foot tower.

Getting reliability means spending real dollars upfront. One colleague says, "Second-time wind generator buyers want the most expensive machine on the market." Although cost isn't always a direct indicator of quality, you do need to pay the big bucks to get high reliability. Do it. (I provide additional guidance on different parts of a wind system in Part III.)

Install Your System Expertly

You can buy the best equipment on the market, do everything right in your design, and still end up wasting your money. To make good on your wind energy investment, you need to install all the equipment well.

Installation takes quite a number of skills (see Chapters 17 and 18), and you'll likely want help (Chapter 12 shows you how to find it). Cutting corners on developing the skills and getting the help decreases your chances of a high-quality installation.

Pay particular attention to tower installation — you don't want it coming down. The details of the electrical installation can also be devilish, and you'll likely need a mentor at a minimum, if not an electrician. Wind-electric systems are a poor place to imitate the ridiculously complex cartoons of Rube Goldberg. The physical forces involved tend to quickly weed out the weak points in design and installation, and your pocketbook, energy balance, and ego will suffer.

Maintain Your System

After you've done an excellent job of designing and installing your wind-electric system, don't lose focus. These systems require regular awareness and periodic maintenance. Although I know of people who've lucked out with machines that keep running with no maintenance, they're clearly the exception, not the rule. I'd go for improving your odds of success by doing a great job of maintenance and keeping it up for the life of the system.

In general, I recommend climbing or tilting down your tower at least once a year. Make that twice a year if you live on a very windy site.

Chapter 21

Ten Wind-Energy Mistakes

*I*f you ignore the mistakes of past wind energy users, you're likely to repeat them. And I see this happen over and over again with people and companies new to the industry. If you're at all interested in wind energy, dig deep and find out what has *not* worked for others. Seeking out others' mistakes is a fertile field for improving your own approach. Ten of the biggest mistakes you can learn from are in this chapter.

Running Afoul of Neighbors and Authorities

You may need to overcome substantial legal hurdles to build and hook up your wind-electric system (see Chapter 2 for details). If you live way out in the country, you may not be subject to such bureaucracy, or you may be able to ignore it; the closer you get to town, the harder that is.

If you don't work well with forceful people in government offices, enlist a friend or hire someone to do the bureaucracy surfing. If you choose to ignore the local requirements, you run the risk of stop-work orders, fines, or perhaps even the disassembly of your system.

I consider getting along with your neighbors to be even more important than appeasing government representatives. Government bureaucrats tend to get out of your hair after you've jumped through all the necessary hoops, but neighbors stick around, so you want to be on good terms with them for the long haul. Here are some good strategies:

> ✔ **Involve your neighbors very early in your dreaming and scheming process.** Seek out their questions and objections and share your

information and perspective. Your best bet is to take your neighbors on a tour of nearby systems so they can get a sense of what you're planning and what it will look and sound like.

✔ **When your system is done, throw a neighborhood party.** Invite your renewable energy friends, installer, and neighbors to celebrate and understand your achievement. And if you have battery backup, invite friends over during utility outages — for promoting the value of renewable backup systems, there's nothing like it!

Underestimating or Overestimating Your Energy Use

If you underestimate your energy use, you may end up not designing a large enough wind energy system to cover your electricity needs. If you overestimate your energy use, you'll spend more money and make more energy than you need.

I suggest that you don't *estimate* your energy use at all — measure it! Knowing how much energy you actually need is the basis of a good home energy design. If it's not in an exact number of kilowatt-hours (kWh) per day, month, or year, do some more homework. Flip to Chapter 6 for the information you need.

Overestimating Your Wind Resource

Your energy use number (kWh) gives you a goal to aim for. Your average wind speed gives you the means to reach that goal and helps determine how much money you need to invest to get there. Although you may need to do some educated estimation here (not really guessing), be very careful and conservative. Use all resources available to determine an average wind speed at hub height, and take the lower numbers. If you underestimate slightly, you'll end up with more energy than you need, which isn't the worst problem to have. Chapter 8 has the details on determining your site's wind energy potential.

Using Too Small a Rotor

A grape doesn't satisfy as many hungry guests as a watermelon does, and a thimbleful of cider doesn't quench your thirst like a mug full does. If you want to catch significant amounts of wind energy, you need a big collector on your wind generator. Home-scale wind turbines, realistically, are 12 to 50 feet in diameter. Smaller wind generators make small amounts of energy. Chapter 13 has information on different wind generator sizes.

Having a Light-Duty Turbine on a Heavy-Duty Site

If you buy a lightweight, high-speed wind turbine, don't expect it to last long, especially if it's in a severe environment. Your grandmother's silver may be passed down to your grandchildren's grandchildren; a plastic fork will likely break before it's fed you two meals. In the same way, a light- to medium-duty wind generator on a medium to heavy-duty site will disappoint you.

Gauging what duty of turbine your potential site needs is somewhat difficult — neither factor comes with labels telling you it's heavy, medium, or light duty. I consider sites with more than a 10 mph average and peaks above 60 mph to be more than medium duty. If you live with a 7 mph average and never see 50, you may be able to consider some of the lighter equipment available. Deciding what is and isn't a heavy duty wind turbine is harder. Weight plays into it, but quality of design and construction is even more important. Look for equipment that has stood the test of time for years in the field. Chapter 16 has more information on specifying a high-quality system.

Buying a Wind Generator without Customer Support

In North America, even though you can certainly find hundreds of wind generators you can order if you have your credit card handy, I'd recommend only a handful of them. That's because I know from years of experience that wind generators almost always need manufacturer support (in terms of warranty, parts, and expertise). I recommend only wind generators that have that support and that have it in your own country. If you build your own wind generator, you are the factory, and you provide the support. But if you buy a manufactured machine, it should come with support. Period.

Using Too Short a Tower

Productive wind generators live on tall towers — this idea is almost impossible to overemphasize. No matter how many people try to bend and break this principle, it's just physical reality that there's more wind the farther you get away from obstructions. The appropriate place to capture meaningful amounts of wind energy is on tall towers. I explain the importance of having a tall tower in detail in Chapter 14.

Considering Only Upfront Costs

If you buy a house and plan only for the market price of the house, you may end up 30 years later with a dilapidated shell of a house and a big pile of debt. You need to consider costs such as financing, maintenance, repairs, and the like. It's no different with a wind-electric system. When you make your financial plans and calculate your financial return, you must look at *life-cycle* cost, not just upfront cost. The full cost includes not only the dollars you pay for maintenance and repair but also ongoing fees associated with operating the system, plus any time you put into it.

Using "Creative" Designs or Parts

Wind energy seems to be a magnet for creativity. And if you're a renewable energy homebrewer with a preference for trying out your wild ideas on yourself, go for it! (Just make sure you know how to avoid shocks and falls and keep your equipment safe.)

The situation gets dicey when experimenters and inventors want to try their ideas out on others, with other people's money — maybe even yours. Testing innovative ideas can be exciting, but it may be a recipe for disaster. I recommend that you avoid the following:

- ✔ Products without a track record (at least three years) and a warranty
- ✔ "Unique" products — good design principles lead to common designs more often than wildly unusual designs
- ✔ Unconventional towers using buildings, trees, or elaborate mechanisms
- ✔ Wind generator or tower designs that you can find from only one source

Ignoring System Maintenance

Avoiding or ignoring the necessary maintenance on your system can lead to gradual wearing out of your equipment. It can lead to dramatic mechanical failure. It can lead to shock and fire. And in all cases, it does lead to systems that suffer performance problems and end up being more expensive in the end. Figure out whatever you need to get your system on a maintenance schedule! Chapter 19 has more info.

Chapter 22

Ten Tales of Wind-Energy Users and "Abusers"

. . . .

. . . .

*W*ind-energy users are pretty hard to pigeonhole. They span the range from multimillionaires to poor farmers and from techie nerds to folks who don't know a volt from an amp (though I hope they know a watt-hour from a watt). Looking at the tales of a variety of wind-energy users and "abusers" can be a useful method for understanding where wind energy might fit in your life. (My tongue's in my cheek with the "abusers," and I include myself as one.) Each story in this chapter raises questions and gives answers about wind energy.

Frank and Deb: Reducing the Propane Bill

System type: Off-grid

Wind generator: Abundant Renewable Energy 110 (12 feet in diameter)

Tower type and height: 168-foot guyed lattice

Wind resource: Estimated 8.5 mph average

Energy yield: Estimated 5.5 kWh per day

Frank and Deb bought an off-grid property with a solar-electric/propane generator hybrid. Their modern home's load required more use of the generator in winter than they wanted, so they invested in a 12-foot diameter wind generator on a tall tower to complement their 3.6 kW solar (PV) array.

Their system has not been without issues — the wind generator is down for repair as I write this. But it has been a strong producer and has reduced their propane bill substantially. Their system has been a test site for a maximum power point tracking (MPPT) charge controller, so not only are they generating electricity, but they're also helping the industry improve its products. Frank and Deb's story shows that off-grid systems often need multiple generating sources and that installing a wind system high and installing it large yields real energy.

John and Lisa: Going On-grid

System type: Off-grid and now on-grid

Wind generator: African Wind Power 3.6 (12 feet in diameter)

Tower type and height: 168-foot freestanding lattice

Wind resource: 10 mph average

Energy yield: Estimated 6 kWh per day

John and Lisa have owned an off-grid property for many years, and they recently began developing it. They initially installed an off-grid wind- and solar-electric system, with the solar array and the wind turbine on the same tower. One major investment in a freestanding tower got them into two resources — sun and wind — neither of which is available at ground level.

Over several years of part-time use of their property and system, they noted that their surplus energy was being wasted when their batteries were full, and the couple decided to connect to the grid. While retaining their goal of a zero-energy home, today whenever the wind and sun are providing more energy than they need, they get a credit from the local utility. John and Lisa's experience illustrates how a properly designed wind generator tower can double as a solar-electric array platform. It also shows that the utility grid is a great "battery," absorbing surplus energy and allowing smaller battery banks.

Doug and Alicia: Facing Maintenance Problems

System type: Battery-based on-grid

Wind generators: Whisper 3000 and then Jacobs (16 feet in diameter)

Tower type and height: 110-foot guyed lattice

Wind resource: Estimated 8 mph average

Energy yield: Estimated 9 kWh per day

Doug and Alicia are private folks who retired from the high-tech world to a stunning view property in a gated community. Before the developer provided the promised utilities, Doug bought a wind generator and then found a local contractor to help install it. Minor problems with that turbine led to a second turbine. Major problems with it led to replacement.

Doug and Alicia were extremely patient through the many problems and failures, which a variety of contractors addressed with various levels of success. In the end, a dump truck damaged their tower by hitting a guy wire, and Doug and Alicia decided that their patience for wind energy had run out. They sold their last wind turbine, which is now running in another location with a hands-on owner who's able and willing to deal with its peculiarities. Meanwhile, their passively tracked solar-electric array continues to produce a portion of their electricity, which is used in their home or sold back to the local utility. Doug and Alicia's problems underscore a couple of important points about wind energy: One, it's not for everyone, and two, buying medium-duty gear without a sufficient track record is a gamble.

Dean and Betty: Estimating Resources

System type: Batteryless grid-tie

Wind generator: Bergey Excel (21 feet in diameter)

Tower type and height: 100-foot freestanding lattice

Wind resource: 4.3 mph average

Energy yield: 1.4 kWh per day

Dean and Beatty got excited about an incentive program in their area and decided to put in a wind generator. The cost of *anemometry* (wind speed measurement) deterred them from measuring their wind, so they went by their gut, which turned out to be very wrong. Their site ended up having a very poor wind resource, despite its topography, and their $55,000 investment was not rewarded with much energy production. They were able to sell the system (but not the 48 yards of concrete in the ground), recouping a portion of their investment. Today, using the improved wind maps would've avoided this costly mistake.

Dean and Betty's mistake stresses the vital point that you have to know your resource or prepare for possible regret! Flip to Chapter 8 for information on determining your site's wind energy potential.

Randy and Melissa Richmond: Giving a System a New Home

System type: Batteryless grid-tie

Wind generator: Bergey Excel (23 feet in diameter)

Tower type and height: 100-foot freestanding lattice

Wind resource: 12.5 mph average

Energy yield: 44 kWh per day

Randy and Melissa bought their system from Dean and Betty (see the preceding section). They got a great deal on a not-very-used system. In contrast to Dean and Betty's site, Randy and Melissa have an excellent wind resource. Two wind farms are in their general area, and their specific site is open, with no trees, and good exposure to the prevailing wind.

Though the system has been operational only half a year, all signs indicate that it will be very productive in its new home, and Randy is well equipped to maintain the system. He's a hands-on guy who's into data nerding, and his small software company (www.righthandeng.com) is contributing to the pool of equipment for keeping track of systems. Randy's good fortune shows that a good resource plus a large rotor equals a lot of kilowatt-hours.

Hugh Piggott: Building His Own

System type: Off-grid

Wind generator: African Wind Power 3.6 m (12 feet in diameter)

Tower type and height: 70-foot guyed tilt-up tower

Wind resource: 12.8 mph average

Energy yield: 7 kWh per day

Hugh is one of my earliest wind energy gurus, and in fact, he's the technical reviewer for this book. He lives on a remote, off-grid peninsula in northwest Scotland. He was determined to have electricity for his family about 30 years ago, so he began making wind generators. Today, his neighborhood is full of home-built wind generators, mostly of his design. He lives with wind and solar electricity and solar hot water in a super energy-efficient house he built.

Hugh started from scratch and with *no* energy, and has gradually built his system up to a point where it provides most of his electricity. This build-as-you-go philosophy works well for some, and if you end up with too much energy, as Hugh says, "It's not hard to use the stuff."

Hugh sells books and plans on how to build your own wind generator, and he teaches courses in Scotland, in Northwest Washington with me, and in other world locations. You can find out more about his work at `www.scoraigwind.com`. Hugh has helped thousands of people learn how to build their own wind generators; his success shows that building your own equipment as you can afford it is a workable solution if you have a homebrew temperament and great persistence.

The Dans: Sharing Innovation

System type: Off-grid

Wind generators: Homebuilt otherpower.com axial field machines (7 feet in diameter and 20 feet in diameter)

Tower type and height: 45- and 85-foot tilt-up towers

Wind resource: 8.5 and 10 mph averages

Energy yield: 1 kWh per day and 15 kWh per day

Dan Fink and Dan Bartmann — "the Dans" to me and others — live in an off-grid neighborhood in the northern Colorado mountains. They learned about home-built wind generators from Hugh Piggott and have built nearly 200 machines for their friends, neighbors, and clients.

These creative guys have had many exciting failures, and the Dans continue to learn from them, sharing their lessons with wind power enthusiasts worldwide at the seminars they teach, in their book *Homebrew Wind Power* (Buckville Publications LLC), and online with their Web site (`www.otherpower.com`) and discussion board (`www.fieldlines.com`). Their story highlights how sharing your enthusiasm for wind energy breeds converts. Hugh's excitement excited the Dans, who now excite others.

Robert Preus: Being a True Professional

System type: Batteryless grid-tied

Wind generators: ARE 442 and 110 (24 feet in diameter and 12 feet in diameter)

Tower type and height: 143-foot freestanding lattice and 127-foot tubular tilt-up towers

Wind resource: 12.3 mph average

Energy yield: 53 kWh per day and 13 kWh per day

Robert started his career in wind doing forensics in the utility and small commercial-scale sector; his job was to understand specifically how and why wind generators fail. After many years in this role, Robert decided to get his feet wet in the home-scale wind energy field by learning how to rebuild direct-drive Jacobs generators at the elbow of wind-energy expert Mick Sagrillo and by importing the African Wind Power 3.6 meter machine from Zimbabwe.

After experiencing problems with the quality and business practices of AWP, Robert decided to build his own line of wind generators with his company, Abundant Renewable Energy (www.abundantre.com), roughly modeled after the original Hugh Piggott design used by AWP but significantly improved. Though his business has had struggles, the company continues to produce high-quality machines, and his deep understanding of what makes wind generators survive or fail has served him well.

Robert lives with both models of his ARE turbines in a serious wind resource, so he knows how they work on a day-to-day basis. His admirable business ethics lead him to be transparent about problems with his machines, and to help installers and users resolve issues. His example shows that attention to what makes wind generators, business, and customer service work can lead to productive installations.

Highland Energy: Sharing Cell Towers

System type: Batteryless grid-tie

Wind generators: Proven 6 and preparing for Proven 15 (18 feet in diameter and 29 feet in diameter)

Tower type and height: 170-foot freestanding lattice

Wind resource: Estimated 9 mph average

Energy yield: Estimated 3 kWh per day

Highland Energy is a cell tower and equipment installation and maintenance company that works all over the western United States and beyond. A few years ago, it decided to get into the renewable energy business, with the stated goal of putting renewables on as many cell towers as possible.

The company wisely started with its own tower at its headquarters and has found that the small-wind industry is not as mature as the cell industry. The purchasing department at Highland has discovered that the quick ordering lead times it's used to in its fast-paced industry are not at all the norm in the small-wind industry. Highland's first wind generator was a long time being delivered, and it then gave its owners a few years of problems. Highland is about to install a larger (and it hopes more reliable) model, and it remains committed to renewable energy's role in its business model. Highland's experience suggests that experienced contracting professionals can move into wind energy, but they will have growing pains and need determination to make it work.

Yours Truly: Always Experimenting

System type: Off-grid

Wind generators: Presently African Wind Power 3.6, Bergey XL.1, Whisper 200 rated at 1,000 W (12 feet in diameter, 8 feet in diameter, and 9 feet in diameter)

Tower type and heights: 112-foot, 150-foot, and 150-foot homebuilt fixed guyed towers

Wind resource: 7.5 mph average

Energy yield: Estimated 9 kWh per day average when all are working

In a classic example of "do as I say, not as I do," I'm an experimenter and not a great maintainer of wind generators. I run two or three machines at a time because I like to test and play with them and see what I can learn. I also have one tower that's clearly too short and two that could be taller. (Find out all about the importance of tall towers in Chapter 14.)

My family and I live on a site with very tall trees (a few top 140 feet, though most are between 80 and 120). If energy production were my only motivation, I'd install one very tall (220-foot?) tower and one very large wind generator. Instead, I play with smaller machines, learn lessons to share with you and others, and dream of a taller, conventional tower. My lessons help others avoid mistakes, and our off-grid homestead doesn't use much propane for the backup generator in the winter because our hybrid off-grid system takes care of most of our load most of the time. Lessons: Wind energy readers should be careful about following "experts'" examples but can definitely learn from their mistakes.

Part VI
Appendixes

The 5th Wave   By Rich Tennant

"Take it easy, everyone. Let's just hope the wind currents carry this thing out of here."

In this part . . .

This part is where I help you understand the nerdy terminology and numbers related to wind energy. You find a brief glossary of wind electricity terms, conversions for size and speed, and assorted abbreviations and acronyms. This reference material can come in handy as you continue to study wind energy.

Appendix A

Glossary

air density: Mass per unit volume of air; air is less dense by roughly 3 percent for every 1,000 feet of elevation above sea level

airfoil: A blade shaped to optimize the lift/drag ratio and to maximize the wind generator's energy production

alternating current (AC): Charge (electron) flow in two directions

alternator: A rotating device that generates AC electricity by passing magnetism past coils of wire

ammeter: A device that measures amperage

ampacity: The maximum safe amperage-carrying capacity of wire or an electrical device

amperage: The rate of charge flow in electrical circuits, measured in *amperes,* or *amps* (A); popularly called *current*

amp-hour (Ah): Unit of charge; used for specifying battery capacity

anemometer: A device that measures wind speed

annual energy output (AEO): The amount of energy in kilowatt-hours that a specific wind generator will produce at a specific average wind speed over a year

average wind speed: The average of available wind in a set period, typically one year

balance of systems (BOS): Wind-electric system components besides the wind generator and tower, typically including the charge controller, voltage clamp, batteries, inverter, disconnects, overcurrent protection, grounding, and so on

base: The concrete foundation that supports the steel structure of a wind generator tower

battery: A group of electrochemical cells that store electrical energy via chemical reactions

batteryless grid-tied system: A wind-electric system that connects to the utility grid without batteries, providing no backup for utility outages

Betz limit: The maximum percentage of wind energy (about 60 percent) that a perfect wind generator could capture

Blade: An airfoil designed for capturing wind energy

blade pitch control: A method of governing that twists the orientation of the blades to degrade the airfoil and spill energy in high-wind conditions

brake switch: A switch that allows the owner to electrically brake (stop the machine through short-circuiting the three phases) a wind generator

breaker: A device to allow the disconnection of electrical devices; it also provides overcurrent protection

brushes: Devices that, along with slip rings, allow transmission of electricity from a rotating or yawing portion of a wind generator to a fixed portion

capacity factor: The ratio of actual energy generated to the potential energy that would be generated at peak rated output at constant peak rated wind speed, expressed as a percentage; this utility-scale measure speaks to wind resource as well as turbine efficiency

carabiner: A steel or aluminum connector for attaching gear to a harness, tower, lines, and so on

charge: The moving material that defines electricity; charged particles; electrons when in wires and ions elsewhere

charge controller: An electronic device that prevents battery overcharge; other common functions include maximum power point tracking (MPPT), step-down, and undercharge protection; usually used with a dump load (*see also* dump load; maximum power point tracking; step-down)

circuit: An electrical pathway for charges; a complete loop must be present for electrical energy to travel (*see also* parallel; series)

compact fluorescent light (CFL): A fluorescent light bulb with built-in electronics and typically with a standard threaded base; it uses $1/4$ the energy of an incandescent bulb that produces the same amount of light

conduit: Metal or plastic pipe to carry electrical wires

cost of energy (COE): Cost in cents per kilowatt-hour of electrical energy

cube law: Wind power is proportional to the cube of the wind speed ($V \times V \times V$, or V^3)

current: A common-language term for the rate of charge (electron) flow; also known as *amperage*

cut-in: The wind speed at which a wind generator starts to generate, typically 5 to 9 mph, below which there is very little energy

cut-out: The wind speed at which a wind generator stops producing, typically not applicable with home-scale turbines, which may or may not continue generating in high winds (*see also* governing)

depth of discharge (DOD): Level of battery discharge expressed in the percentage of charge removed; it's the inverse of *state of charge*

digital multimeter (DMM): A device that measures voltage, amperage, resistance, and possibly other electrical qualities

direct current (DC): Flow of charge in one direction

direct drive: A type of wind turbine with blades connected directly to the generator, with no gears, pulleys, or belts

disconnects: Mechanical means for disconnecting electrical devices from each other, such as loads on circuits from energy sources

downwind: Describes a wind turbine with rotor (blades) on the lee or downwind side of the tower

D-ring: A *D*-shaped attachment point on a climbing harness; typically there are two at the hip, two at the waist, one at the chest center, and one at the back center

dump load: An air or water heater that dissipates excess energy from a wind turbine, usually controlled by the charge controller

efficiency: The ratio of energy out to energy in, expressed as a percentage

energy: Work done over time; power (watts) $\times$ time (hours) = energy (watt-hours)

energy curve: A graphic presentation of the energy (watt-hours) produced by a wind generator in a range of specific average wind speeds

fall-arrest system: A system to prevent tower climbers from falling in the event of fainting, slipping, or having other accidents

fixed guyed tower: A lattice or tubular tower supported with guy wires

flagging: Deformation of trees caused by the wind (*see also* Griggs-Putnam index)

freestanding tower: A lattice or tubular tower supported by concrete and steel, with no guy wires

frequency: The number of times per second that alternating current (AC) electricity changes direction twice for a complete cycle; measured in hertz (Hz)

furling: A form of governing that reduces the exposure of an airfoil to winds to protect the machine in high winds

gear driven: A type of wind turbine that has gears and shafts between the rotor (blades) and generator

gin pole: A pole at right angles to the tower used as a lever to raise tilt-up towers; also, a vertical pole temporarily attached on a non-tiltable tower to lift the tower sections, wind generator, or parts

governing: Protection from excessive rotor speed in high winds; done through techniques such as furling, blade pitch, or electrical braking

grid: Utility lines and infrastructure

grid-tied: A wind-electric system connected to the utility grid; may or may not have batteries for outage protection

Griggs-Putnam index: A system that correlates the amount of tree flagging (deformation) with the wind resource

grounding: A method of connecting all metal devices in a system to each other and to the earth to reduce lightning vulnerability and to shunt (divert) fault current to the ground

guy anchor: A steel and/or concrete device to hold guy wires for tilt-up and fixed guyed towers

guy wire: Steel cable used to support guyed towers

harness: A webbing-and-steel climbing apparatus worn for tower work

hertz (Hz): The unit of frequency of alternating current (AC) electricity (*see also* frequency)

horizontal-axis wind turbine (HAWT): A wind generator that spins on a horizontal shaft, like most available wind turbines do today

hub: The center of a wind generator's rotor, where the blades are connected to each other and to the main alternator shaft

hub height: The height at the center of the wind generator rotor

incentives: Financial subsidies and policies that encourage installation of renewable energy systems

induction generator: A brushless, nonpermanent magnet motor that's spun faster than its normal speed to generate electricity; magnetism is induced in the rotor by the shifting magnetism in the stationary portion of the generator

instantaneous wind speed: Wind speed at a given moment

inverter: A device that converts DC electricity to (usually high-voltage) AC

kilowatt (kW): Rate of energy generation, transfer, or use

kilowatt-hour (kWh): Unit of electrical energy

lanyard: A fixed or adjustable line that attaches climbers or gear to a tower

light emitting diode (LED): A robust electronic device that creates light very efficiently

load: Electrical appliances, either specifically or combined, as in a home's total energy load

maximum power point tracking (MPPT): Adjusting the operating voltage of a wind generator at different speeds to maximize the output of a wind turbine and capture the most energy possible

net metering: A system of utility interconnection that credits a home renewable energy system for kilowatt-hours generated at the same rate the utility charges, up to the level of the home's usage

nose cone: Fiberglass or metal cowling in front of a wind generator's rotor

off-grid system: A wind-electric system not connected to the utility grid

ohm (K): The unit of electrical resistance

Ohm's Law: An electrical law that states that amperage is equal to voltage divided by resistance (A = V ÷ K)

overcurrent protection: Electrical protection against high amperage in a circuit, which may threaten to melt wires and cause fires; typically consists of circuit breakers or fuses

parallel: Connecting electrical devices in parallel paths in a circuit, with all positives and all negatives wired together; it increases amperage or amp-hour capacity in batteries while voltage remains the same

payback: The number of years energy-generating or energy-saving equipment takes to pay back its original investment

peak power: Maximum wattage that a device generates

peak wind speed: Maximum wind speed that a device experiences

permanent magnet alternator (PMA): An electricity-generating device that has permanent magnets (instead of electromagnets) passing by coils of wire to generate electricity

phantom load: An electrical appliance that still uses energy when switched off, due to poor appliance design or needed functionality while off

photovoltaic (PV): Solar-electric; generating electricity with sunlight

power: Rate of energy generation, transmission, or use; wattage

power conditioning equipment: Equipment such as rectifiers, controllers, inverters, and voltage clamps that change incoming energy from a wind generator so that it's compatible with the grid or other equipment and to properly charge batteries

power curve: A curve that plots the instantaneous wattage of a wind generator across the range of wind speeds

power factor: The percentage of amperage that's aligned with voltage and contributes useful power in a circuit

power formula: In electrical terms, power = voltage × amperage × power factor; in wind terms, power = $\frac{1}{2}$ air density × swept area × wind speed cubed

rated power: The manufacturer's claim of output at the rated wind speed, often the typical peak

rated wind speed: The wind speed at which a wind generator's rated power output is measured

rectifier: A device that converts AC electricity to DC

renewable energy (RE): Energy generated from wind, sun, falling water, and biomass (plant and animal materials); it recurs regularly through natural cycles and is less damaging to the environment than fossil fuels

resistance (electrical): A measure of how much an electrical device opposes the flow of charge

resource: Wind (or other renewable) energy available

rotor: The blades and hub of a wind generator; also the rotating portion of an alternator

series: Connecting devices in a circuit in a daisy-chain fashion: positive to negative to positive; it increases voltage while maintaining amperage or amp-hour capacity in batteries

sine wave: A smooth, alternating current (AC) wave

slip rings: Rings made of brass that graphite brushes press against to make a sliding contact, allowing transmission of electricity from a rotating or yawing portion of a wind generator to a fixed portion

start-up: The wind speed at which a wind generator starts spinning, typically somewhat before cut-in (when the generator starts producing electricity)

state of charge (SOC): The measure of capacity of battery remaining, as a percentage; it's the inverse of _depth of discharge_

stator: The stationary portion of an alternator

step-down: Adjusting from higher to lower voltage in an electrical device

swept area: The area (in square feet or square meters) that a wind generator's blades sweep; equal to pi times the blade rotor's radius squared (πr^2)

tag line: A line used to control items being raised and lowered on a tower

tail boom: A metal tube or rod that extends behind the wind generator and holds the tail vane on the wind generator

tail vane: The sheet of metal, wood, or plastic on tail boom that directs turbine rotor into wind

30/500 rule: A guideline that prescribes that a wind generator's lowest blade tip should be a minimum of 30 feet above anything within 500 feet to get into a quality wind resource and minimize destructive turbulence

tilt-up tower: A tower that's typically not climbable but can be lowered to the ground using cables, lifting gear, and a gin pole (*see also* gin pole)

tower adaptor: A welded or cast piece that connects the wind generator to its tower

transmission: Wires used to transmit electricity from the wind generator to batteries, loads, and/or the grid

upwind: Describes a wind turbine with rotor (blades) on the windward side of the tower

utility interconnection equipment: Disconnects, meters, and wiring that connect the inverter to the utility grid, for the purpose of "selling" wind energy to the utility

vertical-axis wind turbine (VAWT): A wind generator that spins on a vertical shaft, unlike most machines available today

volt (V): The unit of electrical pressure

watt (W): A unit of *power* (wattage), the rate of energy generation, transmission, or use

watt-hour (Wh): The unit of electrical energy

wind generator: A device for generating electricity using wind as the motive force; also called a *wind turbine*

wind rose: A graphic presentation of which directions a specific site's wind comes from

wind shear: The rate of increase in wind speed as you move away from the ground

wound field alternator: An alternator that uses wound coils as electromagnets to produce the field magnetism

yaw: The motion of a wind generator around the tower to face the wind

yaw bearing: The bearing (typically a ball bearing) that allows a wind generator to turn and face the wind

Appendix B

Conversions, Abbreviations, and Acronyms

● ●

*M*any North Americans live outside the metric world. But wind-energy systems, especially those from Europe and larger scale systems, often use metric terms. The conversions here can help make the transition between the two systems.

This appendix also includes some abbreviations and acronyms to help you understand wind-energy and electricity shorthand.

Size Conversions

1 inch = 2.54 cm

1 cm = 0.39 inches

1 foot = 0.305 meters

1 meter = 3.28 feet

1 square foot = 0.093 square meters

1 square meter = 10.8 square feet

Speed Conversions

1 mile per hour = 0.446 meters per second = 1.61 kilometers per hour

1 meter per second = 2.24 miles per hour = 3.60 kilometers per hour

1 kilometer per hour = 0.62 miles per hour = 0.28 meters per second

Abbreviations and Acronyms

See the glossary in Appendix A for definitions of many of the following abbreviations and acronyms.

A: Amps

AC: Alternating current

AEO: Annual energy output

Ah: Amp-hours

AWG: American wire gauge

BOS: Balance of systems

CFL: Compact fluorescent light

COE: Cost of energy

DC: Direct current

DMM: Digital multimeter

DOD: Depth of discharge

DOE: Department of Energy

DSIRE: Database of State Incentives for Renewable Energy

EERE: The DOE's Office of Energy Efficiency and Renewable Energy

HAWT: Horizontal-axis wind turbine

Hz: Frequency

K: Ohms

kph: Kilometers per hour

kWh: Kilowatt-hours

LED: Light-emitting diode

mph: Miles per hour

MPPT: Maximum power point tracking

m/s: Meters per second

NOAA: National Oceanographic and Atmospheric Administration

NREL: National Renewable Energy Laboratory

NWS: National Weather Service

PMA: Permanent magnet alternator

PV: Photovoltaic

R: Resistance (measured in ohms)

RE: Renewable energy

SOC: State of charge

V: Volts

VAC: Volts alternating current

VAWT: Vertical-axis wind turbine

VDC: Volts direct current

W: Watts

Wh: Watt-hours

Index

• Y •

• Z •

Business/Accounting & Bookkeeping

Bookkeeping For Dummies
978-0-7645-9848-7

eBay Business
All-in-One For Dummies,
2nd Edition
978-0-470-38536-4

Job Interviews
For Dummies,
3rd Edition
978-0-470-17748-8

Resumes For Dummies,
5th Edition
978-0-470-08037-5

Stock Investing
For Dummies,
3rd Edition
978-0-470-40114-9

Successful Time
Management
For Dummies
978-0-470-29034-7

Computer Hardware

BlackBerry For Dummies,
3rd Edition
978-0-470-45762-7

Computers For Seniors
For Dummies
978-0-470-24055-7

iPhone For Dummies,
2nd Edition
978-0-470-42342-4

Laptops For Dummies,
3rd Edition
978-0-470-27759-1

Macs For Dummies,
10th Edition
978-0-470-27817-8

Cooking & Entertaining

Cooking Basics
For Dummies,
3rd Edition
978-0-7645-7206-7

Wine For Dummies,
4th Edition
978-0-470-04579-4

Diet & Nutrition

Dieting For Dummies,
2nd Edition
978-0-7645-4149-0

Nutrition For Dummies,
4th Edition
978-0-471-79868-2

Weight Training
For Dummies,
3rd Edition
978-0-471-76845-6

Digital Photography

Digital Photography
For Dummies,
6th Edition
978-0-470-25074-7

Photoshop Elements 7
For Dummies
978-0-470-39700-8

Gardening

Gardening Basics
For Dummies
978-0-470-03749-2

Organic Gardening
For Dummies,
2nd Edition
978-0-470-43067-5

Green/Sustainable

Green Building
& Remodeling
For Dummies
978-0-4710-17559-0

Green Cleaning
For Dummies
978-0-470-39106-8

Green IT For Dummies
978-0-470-38688-0

Health

Diabetes For Dummies,
3rd Edition
978-0-470-27086-8

Food Allergies
For Dummies
978-0-470-09584-3

Living Gluten-Free
For Dummies
978-0-471-77383-2

Hobbies/General

Chess For Dummies,
2nd Edition
978-0-7645-8404-6

Drawing For Dummies
978-0-7645-5476-6

Knitting For Dummies,
2nd Edition
978-0-470-28747-7

Organizing For Dummies
978-0-7645-5300-4

SuDoku For Dummies
978-0-470-01892-7

Home Improvement

Energy Efficient Homes
For Dummies
978-0-470-37602-7

Home Theater
For Dummies,
3rd Edition
978-0-470-41189-6

Living the Country Lifestyle
All-in-One For Dummies
978-0-470-43061-3

Solar Power Your Home
For Dummies
978-0-470-17569-9

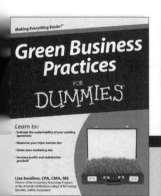

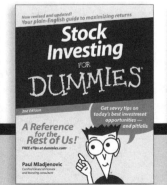

Available wherever books are sold. For more information or to order direct: U.S. customers visit www.dummies.com or call 1-877-762-2974.
U.K. customers visit www.wileyeurope.com or call (0) 1243 843291. Canadian customers visit www.wiley.ca or call 1-800-567-4797.

Internet

Blogging For Dummies,
2nd Edition
978-0-470-23017-6

eBay For Dummies,
6th Edition
978-0-470-49741-8

Facebook For Dummies
978-0-470-26273-3

Google Blogger
For Dummies
978-0-470-40742-4

Web Marketing
For Dummies,
2nd Edition
978-0-470-37181-7

WordPress For Dummies,
2nd Edition
978-0-470-40296-2

Language & Foreign Language

French For Dummies
978-0-7645-5193-2

Italian Phrases
For Dummies
978-0-7645-7203-6

Spanish For Dummies
978-0-7645-5194-9

Spanish For Dummies,
Audio Set
978-0-470-09585-0

Macintosh

Mac OS X Snow Leopard
For Dummies
978-0-470-43543-4

Math & Science

Algebra I For Dummies
978-0-7645-5325-7

Biology For Dummies
978-0-7645-5326-4

Calculus For Dummies
978-0-7645-2498-1

Chemistry For Dummies
978-0-7645-5430-8

Microsoft Office

Excel 2007 For Dummies
978-0-470-03737-9

Office 2007 All-in-One
Desk Reference
For Dummies
978-0-471-78279-7

Music

Guitar For Dummies,
2nd Edition
978-0-7645-9904-0

iPod & iTunes
For Dummies,
6th Edition
978-0-470-39062-7

Piano Exercises
For Dummies
978-0-470-38765-8

Parenting & Education

Parenting For Dummies,
2nd Edition
978-0-7645-5418-6

Type 1 Diabetes
For Dummies
978-0-470-17811-9

Pets

Cats For Dummies,
2nd Edition
978-0-7645-5275-5

Dog Training For Dummies,
2nd Edition
978-0-7645-8418-3

Puppies For Dummies,
2nd Edition
978-0-470-03717-1

Religion & Inspiration

The Bible For Dummies
978-0-7645-5296-0

Catholicism For Dummies
978-0-7645-5391-2

Women in the Bible
For Dummies
978-0-7645-8475-6

Self-Help & Relationship

Anger Management
For Dummies
978-0-470-03715-7

Overcoming Anxiety
For Dummies
978-0-7645-5447-6

Sports

Baseball For Dummies,
3rd Edition
978-0-7645-7537-2

Basketball For Dummies,
2nd Edition
978-0-7645-5248-9

Golf For Dummies,
3rd Edition
978-0-471-76871-5

Web Development

Web Design All-in-One
For Dummies
978-0-470-41796-6

Windows Vista

Windows Vista
For Dummies
978-0-471-75421-3

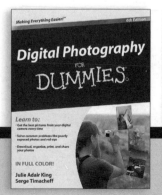